The Human Weapon System

Christopher Myers

The Human Weapon System

Springer

Christopher Myers
3 Sports Science
Founder and Head Coach
Akron, OH, USA

ISBN 978-3-031-45060-0 ISBN 978-3-031-45061-7 (eBook)
https://doi.org/10.1007/978-3-031-45061-7

This Springer imprint is published by the registered company Springer Nature Switzerland AG
The registered company address is: Gewerbestrasse 11, 6330 Cham, Switzerland

Paper in this product is recyclable.

To the men and women serving on the front lines domestically and internationally in order to keep us safe, Thank You.

Introduction

In a world where excellence is no longer a luxury but a necessity, the drive to improve human performance for the Human Weapon System (HWS) has never been more crucial. Recognizing this demand for efficiency, productivity, and peak performance, this book, *The Human Weapon System*, offers a transformative exploration of how organizations can unlock the latent potential in the HWS and optimize human performance at scale.

Our journey begins with understanding the differences between a tactical athlete (TA) and the HWS in Part I. Though related, a TA and the HWS represent distinct concepts in human performance optimization. A tactical athlete refers to military personnel, firefighters, law enforcement officers, or emergency services personnel who rely on diverse physical and mental skills to perform their duties, often under extreme conditions. The development of these individuals focuses on a balance of strength, agility, endurance, and cognitive skills, like professional athletes, but with the added complexity of unpredictable environments and high-stakes outcomes. This term looks at tactical professionals through the lens of an athlete. This section discusses why applying the term TA does not correctly describe these professionals.

On the other hand, the HWS is a holistic framework that views the human body as an integrated system, like a tool or piece of equipment. This concept emphasizes the **interlacing** between physical, mental, and emotional components and aims to optimize the function of the "system" as a whole. This perspective is often used to better conceptualize tactical, law enforcement, firefighter, and emergency services professions. This term focuses on human factors such as nutrition, fitness, mental resilience, and recovery strategies. It is a recognition that performance optimization for tactical athletes requires more than just physical training; it also involves supporting the overall health and well-being of the individual to ensure peak performance in all aspects of their demanding roles.

In Part II, our journey shifts to understanding the concept of human performance optimization (HPO) – the intricate fusion of science, strategy, and practical application to enhance individual and team capabilities. This book aims to elucidate the significance of HPO and explain its pivotal role in driving the success of modern organizations. At its core, this book is a deep dive into the inner workings of an effective HPO program. It presents a careful dissection of the various elements that constitute an HPO program, taking readers beyond mere theory and illuminating the intersection of neuroscience, psychology, management, and technology that lays the foundation for these programs.

My aim is to help leaders, managers, medical professionals, and anyone interested in boosting organizational performance navigate creating and implementing an effective HPO program. This book presents a step-by-step blueprint to create a customized HPO program, drawing from cutting-edge research, case studies, and expert interviews.

Despite the evident importance of HPO programs in enhancing HWS performance, more research in this area still needs to be done. While studies have been on individual components of HPO, like physical fitness, cognitive performance, or emotional resilience, comprehensive, evidence-based research on holistic HPO programs is still nascent. This dearth of research can lead to challenges in developing and implementing effective HPO programs, given the lack of empirical evidence to guide best practices. It also makes measuring these programs' overall effectiveness and return on investment easier. Hence, there is an urgent need for more rigorous and systematic studies in human performance optimization.

Part III will examine if an HWS HPO program can perform periodization as human performance professionals often create and how an HPO program should conduct a programmatic level Needs Analysis. Finally, this section will examine the use of technology to collect and perform data analytics to monitor, analyze, and improve human performance.

The focus of this book is to open the discussion and provide guidance on how to create an HPO program at a programmatic level concerning the HWS professions. This book will create future discussions on the benefits, shortcomings, and makeup of HWS-centric HPO programs. This book will not address other human performance topics such as human physiology, the use of PEDs in HWS professions, and other various subjects on the individualization of human performance. This information can be found in several research manuscripts and textbooks cited throughout this book.

In this book, I aim to catalyze a paradigm shift in your perception of human potential and its optimization. The goal is to move beyond the status quo and embrace a future where people perform at their best, not by accident but by design. Prepare yourself for a journey into the frontiers of human potential as we explore the blueprint to create a culture where every individual HWS can thrive and contribute to the shared organizational vision.

So, get ready to turn the pages, delve into the nuances of human performance optimization, and unlock your team's potential. Welcome to the future of the HWS. Welcome to "The Human Weapon System."

Acknowledgments

My wife and son, thank you for supporting me in this endeavor. Your encouragement and love help make this book possible.

To the 2nd Air Support Operations Squadron (ASOS) leadership and Operators, thank you for showing and teaching me your tradecraft and making me a better Exercise Physiologist.

To Dr. Lea Johansen and Dr. Regina Shia for the long talks on the Cognitive Domain, and to the all the other AFSPECWAR and HP professionals who have shared their knowledge to help grow the human performance initiative within the US Air Force.

To Dr. Elizabeth Foley and Kosta Telegadas for helping write and providing their knowledge and expertise to the Nutrition and Periodization chapters, respectively.

Contents

Part I
Defining the Human Weapon System (HWS) and Human Performance Optimization (HPO)

Chapter 1
The Tactical Athlete

Introduction

In the early 2000s, human performance programs were applied to U.S. Military Special Operations Forces, particularly in strength and conditioning, modeled after professional and Division 1 NCAA athletic programs. The application of these athletic performance programs led to the term "tactical athlete" (TA) being applied to the U.S. Military, police, and first responders [1–5].

Typically, a "tactical athlete" is a term often used to describe professionals who must maintain a high level of physical fitness and mental toughness to perform their jobs. It is generally thought that these individuals typically engage in specialized training regimens that focus on functional fitness, encompassing a variety of exercises and routines designed to improve their overall performance and reduce the risk of injury.

Furthermore, the term is also used broadly to describe anyone who trains like a tactical athlete, adopting similar comprehensive and functional fitness routines even if they do not work in the aforementioned professions.

The "Tactical Athlete"

Since the early 2000s, tactical and first responder professionals have been called "tactical athletes." Schofield and Koundini define a TA as "someone who requires physical preparedness (CPP) on which the technical and tactical skills requisite for the sport of competition are developed" [6]. Generally, this statement means tactical athletes must be capable of performing a wide range of activities that require strength, speed, agility, endurance, and flexibility. They also must be prepared to respond to physically demanding situations at a moment's notice, often under

C. Myers, *The Human Weapon System*,
https://doi.org/10.1007/978-3-031-45061-7_1

high-stress conditions. The NSCA's concept and definition emerged from applying collegiate and professional sports human performance (HP) training models [7]. At its core, "tactical athlete" tries to use the base term "athlete" to a population that does not function as athletes.

An athlete is "a person who is proficient in sports and other forms of physical exercise." Usually, a person envisions a professional or high-performance athlete when discussing the traits of an athlete. This image occurs because professional athletes are at the peak of their physical and genetic performance. For this reason, the military and first responder professions began to apply professional and collegiate human performance models to improve physical performance. Therefore, first responders and warfighters began to be called "tactical athletes" (Table 1.1).

At the time, the application of athletic performance principles was good. However, HP programs have evolved to encompass a focus on more aspects of well-being (i.e., True North and POTFF programs) and access to medical care (i.e., Integrated Occupational Specialty (IOS) teams). Even though these human performance programs are becoming more holistic, many are still viewed through the

Table 1.1 High-performance athlete occupational requirements or traits

High-performance athlete occupational requirements or traits
Win the competition
Season has scheduled "in-season" and "out of season" time periods
Rehabilitation is based on optimal recovery time
Cognitive ability in competition or stressful situations
Innovative
Adaptable
Proficient in designated sport-specific tasks and abilities
Exceeds minimum physical standards for sport
Works well with a team
High self-confidence
Unwaivering determination
High level of intrinsic motivation
Type A personality
Accepts criticism to be better
High levels of resiliency to manage setbacks
Ability to manage stress and anxiety while under pressure

Please note that this table is not an exhaustive list

prism of the tactical athlete lens. This view limits the ability of these HP programs to fully assist the tactical, police, and first responder populations.

As will be discussed throughout this book, human performance and human performance optimization for the first responder and tactical populations are very different from that of athletes. Due to these differences, the term tactical athlete must be updated to fully define the key characteristics and traits of the tactical, police, and first responder professions.

References

1. Nevin J. The tactical athlete: optimising physical preparedness for the demands of combat. Prof Strength Cond. 2017:25–34.
2. Scofield DE, Kardouni JR. The tactical athlete: a product of 21st century strength and conditioning [Online]. www.nsca-scj.com
3. Abt JP, Sell TC, Crawford K, Lovalekar M, Nagai T, Deluzio J, Smalley BW, McGrail MA, Rowe RS, Cardin S, Lephart SM. Warrior model for human performance and injury prevention: eagle tactical athlete program (ETAP) part II. J Spec Oper Med. 2010;10:22–33. https://doi.org/10.55460/42CI-ELHP.
4. Sell TC, Abt JP, Crawford K, Lovalekar M, Nagai T, Deluzio J, Smalley BW, McGrail MA, Rowe RS, Cardin S, Lephart SM. Warrior model for human performance and injury prevention: eagle tactical athlete program (ETAP) part I. J Spec Oper Med. 2010;10:2–21. https://doi.org/10.55460/556O-K7N2.
5. Sax van der Weyden M, Toczko M, Fyock-Martin M, Martin J. Relationship between a maximum plank assessment and fitness, health behaviors, and moods in tactical athletes: an exploratory study. Int J Environ Res Public Health. 2022;19. https://doi.org/10.3390/ijerph191912832.
6. Scofield DE, Kardouni JR. The tactical athlete. Strength Cond J. 2015;37:2–7. https://doi.org/10.1519/SSC.0000000000000149.
7. Alvar B, Sell K, Deuster P. Essentials of tactical strength and conditioning [online]. 1st ed. Human Kinetics https://www.nsca.com/store/product-detail/INV/9781450457309/9781450457309

Chapter 2
The Human Weapon System

The "Human Weapon System"

When a medical professional or exercise physiologist performs a needs analysis (Chap. 9), the outcome for an athlete is significantly different from that for a tactical professional. Tactical and first responder professionals have different skill sets and physiological needs than athletes. However, considering the critical physiological, cognitive, psychological, nutritional, and medical needs of tactical and first responder professionals, these professionals need to be categorized as the "human weapon system." The best way to do this is to look at the difference in characteristics between the HWS and tactical athlete (TA) populations while finally defining the HWS.

Sports and tactical populations are fundamentally different. This book will only exist if the literature explains HP and human performance optimization (HPO) for the HWS. All tactical and first responder training certifications and literature primarily focus on strength and conditioning that apply athlete-specific training theory and practice. For example, the National Strength and Conditioning Association (NSCA) created the Tactical Strength and Conditioning Facilitator (TSAC-F) certification in 2005 [1]. This certification was designed to be the standard for training tactical populations. This certification is governed by the textbook **NSCA's Essentials in Tactical Strength and Conditioning** [1]. The book states that the purpose of the TSAC-F program is "to provide a standardized credential that ensures a high level of professional competence among individuals with the knowledge, skills, and experiences to apply scientific principles to training tactical athletes" [1]. However, the **NSCA's Essentials in Tactical Strength and Conditioning** textbook only focuses on physical training theories and methods for physical training. The book applies traditional athletic physical training models to tactical populations; the certification does not consider the differences between the HWS and athletic populations. To tease out the differences in these populations, look at Table 2.1.

C. Myers, *The Human Weapon System*,
https://doi.org/10.1007/978-3-031-45061-7_2

Table 2.1 The human weapon system (HWS) and high-performing-athlete occupation requirements or traits

HWS occupational requirement or trait	Different	Similar	High performing athlete occupational requirement or trait
Mission success requires life and death decisions	x		Win the competition
"Always in season"	x		Season has scheduled "in-season" and "out of season" time periods
Increased rehabilitation with reduced recovery time	x		Rehabilitation is based on optimal recovery time
Cognitive agility in austere environments and adverse conditions	x	x	Cognitive ability in competition or stressful situations
Innovative		x	Innovative
Adaptable		x	Adaptable
Proficient in multiple "sport specific" tasks and abilities	x	x	Proficient in designated sport-specific tasks and abilities
Exceed minimum physical standards		x	Exceeds minimum physical standards for sport
Self-reliant	x		
Works well with a team		x	Works well with a team
High self-confidence		x	High self-confidence
Unwaivering determination		x	Unwaivering determination
High level of intrinsic motivation		x	High level of intrinsic motivation
Type A personality		x	Type A personality
Accepts criticism to be better		x	Accepts criticism to be better
High levels of resiliency to manage setbacks	x	x	High levels of resiliency to manage setbacks
High levels of resiliency to manage the horrors of combat and/or devastation	x		
Failure is not an option attitude	x		
Moral and ethical, especially in "grey circumstances"	x		
Ability to manage stress and anxiety while under pressure	x	x	Ability to manage stress and anxiety while under pressure
Ability to manage stress and anxiety while under life and death situations	x		

Please note that this table is not an exhaustive list

Shared Traits

The first aspect is investigating traits like the HWS and athlete occupations. As shown in Table 1.1, the HWS and a high-performing athlete share some of the following occupational requirements and traits:

- Rehabilitation from injury *
- Innovative
- Adaptable

- Proficient in sport-specific skills and tasks *
- Works well with a team
- Unwavering determination
- High self-confidence
- High level of intrinsic motivation
- Accepts criticism
- High levels of resiliency *
- Ability to manage stress and anxiety while under pressure *

Note: The "*" denotes subtle differences between these traits

High-performing athletes and the HWS do share many qualities. For example, a high-performing athlete must be able to adapt and overcome challenges during competition through innovation and self-confidence. In competition, an athlete or a team will have a game plan to exploit the opposing team's weaknesses. The method may or may not work. A well-trained athlete or team will adjust the game plan to improve their chances of winning. Another way is for an athlete to work harder in competition to succeed. Let us look at a prime example:

> One of the best examples of a team adapting and a high-performing athlete digging deeper occurred in Super Bowl XXXII in 1998. The Denver Broncos played against the defending champion Green Bay Packers. Up to this point in time, the Broncos were 0-4 in Super Bowl appearances.
>
> Going into the 1997 season, the Broncos changed their traditional playbook and rebuilt their team. The team included Hall of Fame members: John Elway, Shannon Sharpe, Gary Zimmerman, and Terrell Davis. The team's starting lineup for offense and defense had ten free agents and eight draftees. This iteration of the Denver Broncos was young. However, the team's high level of technical skill and innovation saw the team have its longest winning streak in franchise history.
>
> Leading into Super Bowl XXXII, the Denver Broncos were 11-point underdogs to the Green Bay Packers. For most of the game, every team could only get an extensive lead. No matter the changes to their game plans, each team appeared to have an answer. However, this all changed in the middle of the third quarter.
>
> The game was tied in the last few minutes of the third quarter, with the Broncos possessing the ball. The Broncos quarterback, John Elway, was known for orchestrating come-from-behind drives. He could compose passing drives that most defenses could not defend. However, in Super Bowl XXXII, this was different. John Elway changed his tactics and drove the Broncos down the field. Just inside the Green Bay Packers' 10-yard line, John Elway ran for a key first down. This was unusual for him since he was one of the oldest quarterbacks in the NFL at this time. During his scramble, he dove headfirst, flipped, and spun in a circle parallel to the ground. This dangerous move could have cost him his career. However, the "helicopter play" provided a pivotal first down that eventually led the Broncos to score the winning touchdown [2].

The Differences and the Similar Traits

Looking back at the attribute list, some characteristics have an "*." This symbol denotes a subtle difference in these trademarks from what appears on the surface. For reference, these characteristics are listed:

1. Rehabilitation from injury
2. Cognitive agility in highly stressful situations
3. High levels of resiliency
4. Ability to manage stress and anxiety while under pressure
5. Proficient in sport-specific skills and tasks

Rehabilitation is different for the HWS and high-performing athletes. Professional and collegiate athletes have instant access to care and time to recover. This is only sometimes the case for the HWS. For the military, access to care is highly different. In tactical situations, treatment may be delayed due to wartime situations or the difficulty of the terrain. In garrison, military medical facilities are usually understaffed and cannot give immediate specialized care. The military follows a process. The system is designed to be cost-effective and help with access to care. However, the downside is that the process could be more prolonged. Here is an example:

> During a field exercise, US Army Captain (CPT) Andrew McClane twisted his left knee. He experiences significant swelling and pain. To receive proper care, he must follow the following steps:
>
> 1. CPT McClane must go to Sick Call. Sick call is the time frame, usually in the mornings, at the local military treatment facility (MTF) where walk-in medical appointments are accepted. These sessions are typically staffed with military medics, nurses, physician assistants, nurse practitioners, and young military doctors. Depending on the medical professional who examines CPT McClane, the professional may order a set of X-rays or send CPT McClane home with a prescription of ibuprofen. If the medical professional orders X-rays, CPT McClane will likely have to make an appointment on a different day and time to have the X-rays taken.
> 2. CPT McClane would most likely receive an appointment for his X-rays the following day. After the X-rays are taken, he will need to make an appointment to have the X-rays read. Before leaving work, he makes his next appointment for the reading. This follow-up appointment may be a week or two later. For this example, CPT McClane obtained an appointment a week later.
> 3. Going into the follow-up appointment to read the X-rays, CPT McClane has waited 8 days for a prognosis for his injured knee. At the appointment, the medical professional interpreting the X-rays most likely will not the same medical professional, who examined him during his sick call visit. The medical professional states that the X-rays are negative. This result would typically lead to an order for magnetic resonance imaging (MRI); however, military medicine usually does not do this order. Since MRIs are expensive, military medicine usually sends clients to physical therapy (PT) before ordering the MRI.
> 4. After his follow-up appointment, CPT McClane schedules his first PT appointment. Before going any further, it must be pointed out that MTFs are responsible for treating thousands of military service members. This is part of the reason that wait times are very long. This aspect is very true for PT. The military service member may have to wait 30–90 days to see the physical therapist for the first PT appointment. For this example, CPT McClane could get an appointment within 30 days. This timing means CPT McClane cannot receive first-level treatment until 38 days after the injury.
>   (a) The medical professional could order an MRI for CPT McClane. This action depends on the local policies, medical professional experience, and the MTF's capabilities.
> 5. The physical therapist will perform diagnostic tests at the PT appointment to determine the knee stability of CPT McClane and any possible injury. The results of the tests deter-

mine that CPT McClane may have a torn anterior cruciate ligament (ACL) in his left knee. After this diagnosis, two options can occur. The first option is that the physical therapist may order an MRI before prescribing a PT treatment plan. The second option is that the PT chooses to perform a PT treatment plan before ordering the MRI. The most likely possibility is that the physical therapist will order the MRI.

Obtaining an MRI image can be a lengthy process. The quickness of the MRI appointment is based on MRI availability. Most MTFs do not have an MRI; they rely on local medical facilities. Obtaining an MRI image appointment for a facility in the surrounding community takes time. In CPT McClane's case, his MTF does not have an MRI. To get an appointment for an "off-base" MRI, the following steps must occur:

- The physical therapist must order the MRI in the MTF's medical database.
- The order needs to be approved by Tricare, the Department of Defense's (DOD's) medical insurance provider. This process can take up to a week to occur.
- Once Tricare approves the MRI order, CPT McClane can make the appointment at any local MRI lab.
- The time for the MRI appointment to occur depends on MRI availability. This length of time can range from 7 to 60 days. For this example, CPT McClane can obtain an appointment a week after receiving the Tricare authorization.

By the time CPT McClane receives his MRI, at least 45 days have passed since his injury. In the case of athletics, the MRI could occur in as little as 24 hours after the occurrence of the damage.

6. The reading of the MRI usually does not occur at the time of the MRI imaging appointment. In athletics, military, and civilian medicine, the results are interpreted by a specialist and sent to the client's primary care physician. The results will be returned to his MTF, in CPT McClane's case. CPT McClane will only be able to make an appointment at his MTF to know the results of his MRI imaging once the MTF receives the results. The time frame for the receipt and processing of the results varies from facility to facility. For this example, CPT McClane could make his appointment 5 days after his imaging appointment. This action means he still does not know the results of his MRI; he can only make the appointment to get the results. These appointments are not immediate. The appointments are based on medical professional availability. CPT McClane could wait a variable amount of time before his appointment occurs. For this example, CPT McClane can make his appointment within 3 days of his MTF's receipt of his MRI imaging results.
7. CPT McClane learns that his left ACL is torn and needs surgery during his MRI image results appointment to repair the damage. By the time he knows the prognosis of his injury, 53 days have lapsed since his injury. Additionally, CPT McClane may not see the same medical professional during his appointment.

The scenario would continue through surgery to recovery. Obtaining a surgical appointment is highly variable. However, this example highlights the extreme differences between athletics and the military health systems. An athlete could have surgery for the same injury as CPT McClane in as short as 7 days after the injury. In CPT McClane's case, he took 53 days only to learn his prognosis. This appears to be a long wait. Additionally, the time to obtain a surgical appointment needs to be addressed. Once the prognosis is made, patient wait times are about 3–6 weeks. CPT McClane could wait 74–95 days from the day of his initial injury before having surgery to fix his ACL tear. So how does this example compare to real-world reported wait times?

The average wait time documented at Madigan Army Medical Center for orthopedic surgery was 69.1 days [3]. The average MTF wait times range from 45–66 min once checked in to when the patient is seen by a medical professional [4, 5]. The average waiting time to receive an MRI image in the United States is 2–4 weeks [6]. Finally, the average Veteran Affairs, which can be used as a surrogate for military medicine and depends on location, can have an average wait time of 15–30 days before seeing a medical professional for an appointment [7].

Police and first responders do not have to wait as long as a military member for this kind of injury. This aspect is due to private insurance. Police and first responders have private healthcare insurance through their agencies; the military provides healthcare to the service members. Personal insurance allows police and first responders to get healthcare locally. Wait times are significantly less than in the military; however, this aspect does not mean treatment is better. For example, obtaining an orthopedic surgery appointment can take 1.5–3 weeks [8].

For professional and high-level collegiate athletes, access to care is different. Arliani et al. describe an ACL surgery that can occur 1–4 weeks after the initial injury in professional soccer [9] This means a professional athlete would be in rehabilitative training almost 3 months before CPT McClane if they injured their ACLs simultaneously! The average recovery time from an ACL surgery is 6–10 months [10]. The differences in the ability to access care mean CPT McClane cannot perform his military duties for up to 3 months longer than necessary if he had the same level of care as professional athletes. The longer a military service member or first responder is away from these primary duties, their units are not at full capacity and at risk for a lower probability of mission success.

The third aspect, "high levels of resiliency," differs between athletes and the HWS. The DoD defines resiliency as "the ability to withstand, recover, and grow in the face of stressors and changing demands" [11]. Essentially, resiliency is how one can "bounce back" from a stressful situation. Resiliency is an essential component for highly successful elite athletes and the HWS. Yes, the key stressors for the HWS are starkly contrastive and more challenging than those of high-performing athletes.

A few of the key stressors for the HWS are as follows:

- Threats to life through combat [12, 13]
- Residential mobility [12–14]
- Separation from family [14]
- Loss [12–14]
- Reintegration into society from combat [12, 13]

A few everyday athlete stressors are as follows:

- Injury and returning from injury [15–17]
- Competition pressures [15–17]
- Being at one's best during the big game [15–17]
- Competition within the team hierarchy [15–17]
- Issues with proper form and technique [15–17]

First responders, the police, and the military must deal with death and trauma [12, 13, 18]. Exposure to highly stressful situations in austere environments and away from home strains one's mental health and can cause psychological problems (i.e., depression, anxiety, post-traumatic stress disorder (PTSD), anger/aggression, substance abuse, etc.) [13, 15, 18, 19]. Similarly, high levels of mental demands are placed on an elite athlete. Yet these demands cause high levels of anxiety, depression, and substance abuse disorder in elite athletes, as seen with the HWS [20]. Yet the athlete population tends to see higher levels of other psychological disorders, such as eating and body dysmorphic disorders [20, 21]. The bottom line is that athletes and the HWS share a significant overlap in possible mental disorders; however, the root causes differ significantly. The term "tactical athlete" cannot be applied to the military, police, and first responder populations. Moreover, high incidents of eating disorders, and body dysmorphic disorder, are seen.

The "ability to manage stress and anxiety while under pressure" is a shared trait between elite athletes and the HWS. Yet the ending circumstances could not be more different. Looking back at the Denver Broncos example earlier in this chapter, when John Elway chose to run for the pivotal first down, this was a split-second decision during the Superbowl. The stakes could not be higher; the pressure could not be more significant. Yet the consequence of this decision was to win a competition. Athletes must perform under pressure during competitions; however, the competition situations fall within the competition regulations that allow for a degree of prediction. Any in-competition adjustments can be made during pauses in the competition or on the sidelines, and the decisions do not have life-and-death consequences. The consequences of decisions for the HWS are far more significant.

The HWS must be able to make split-second life-and-death decisions, usually in obscure conditions that make it challenging to determine enemy versus bystander. Additionally, these environments are asymmetrical and may not follow any known pattern. For example:

> The author of this book, Dr. Chris Myers, served one of him two combat tours in Operation Iraqi Freedom (OIF) II, where he was a 2d Lieutenant (2LT) in the US Army. During one of his daily visits to a local Iraqi police station in central Baghdad, he surveyed the outer perimeter checkpoint with one of his squad leaders. The local station was embedded in the local community and surrounded by residential apartments. At any point, a sniper or enemy combatant could fire into the police station square and building. While surveying, 2LT Myers saw a red laser positioned on the chest of his squad leader. Immediately, he pushed his squad leader down, and they drew their M4s. 2LT Myers and his squad leader noticed the laser from a window above them. As they were about to shoot, a child opened the curtain to the window, where she shone the pen laser. Immediately, 2LT Myers dropped his weapon and pushed the gun of this squad leader out of firing position. This split-second reaction of 2LT Myers and his squad leader to identify and determine the child as a nonhostile entity avoided the accidental shooting of a non-combatant.

The above example occurred within 2–3 s of the laser being shone on the squad leader. Within this short period of time, 2LT Myers and his squad leader had to locate the threat, determine the viability of the threat, and determine the appropriate response while under the duress of possibly getting severely injured or killed. This type of situation is actual pressure. Any decision the HWS makes must fall within that situation's socially recognized morals, ethics, and legal boundaries.

The final attribute that is similarly shared, but with significant differences, between athletes and the HWS is "proficient in 'sport-specific' skills and tasks." As part of a needs analysis (discussed in Chap. 9), look at the essential skills and tasks an athlete needs to succeed in that sport. Additionally, the skills and functions may only be necessary for a particular position, such as a linebacker or a swimming freestyle sprint specialist. The HWS must be proficient in "career field-specific skills and tasks." This rephrasing of "sport-specific" to "career field-specific" denotes a greater depth of skills and tasks. No matter the HWS career field, these professionals need to be proficient in their career field tasks and others. HWS professionals always "train in depth." This idea means there is an overlap in knowledge for the career fields necessary for the team to succeed. For example:

> During the early 2000s, the US Army Military Police (MP) Corps reorganized some line company structures. At the platoon level, the personnel count was about 38 MPs (not including leadership and specialty positions). This platoon structure allowed for three squads of four teams. Each team consisted of three MPs: a squad or team leader, a driver, and a gunner. This structure allowed the teams to be independent and highly mobile, especially in urban operations. When a team was down an individual, the team would have to borrow another MP from a sister team. The replacement MP could be any rank. This situation meant that an MP, at all times, needed to know the critical tasks to complete the job that was one rank higher than the MP's current rank, and the critical functions of the job two ranks below. This in-depth training allowed the MP units to be highly flexible and adaptable with the limited number of MPs available.
>
> Another example is the US Navy Seals. Before the early 2000s, a SEAL team would be comprised of a couple of two officers, a senior chief petty officer, intelligence personnel, air operations personnel, corpsman, ordinance personnel (aka demolitions and/or sniper), submarine operations personnel, and communications personnel. Cross-training in these different career fields was not an option within these teams. Survival and mission success depended on training depth.
>
> The small team structure is perfect for stealth and quickness. However, its strength is also its weakness. The small team numbers mean the SEAL teams do not have the luxury of multiple personnel for a single position. To mitigate this weakness, SEAL teams ensure that each member knows how to perform the jobs of the other specialty positions. By doing this, the US Navy SEALs have turned a weakness into one of their greatest strengths.

The military, conventional, and special operations have a significant cross-over in job training. This ensures that the unit is still functional and can complete its mission. Similarly, this practice occurs with police and first responders. In athletics, teams can have more significant numbers of athletes, allowing multiple players to fulfill the limited positions on a team. For example, in American football, there are a total of 24 positions on a team. Many NCAA Division 1 football teams have 120–130 student athletes; NFL football teams can have up to 55 players on an active roster. Contrarily, an average US firefighter truck may have four to six professionals. Firefighters, and the US Navy Seals, do not have the luxury of having extra people to fill critical positions when someone is injured; these teams must rely on each other to ensure mission success.

Differences

Until this point, several key similarities and subtle differences between athletes and HWS professionals are discussed. The definition of "tactical athlete" misses the critical differences between the athlete and HWS populations. Some of the key differences between these two populations are as follows:

- Mission success requires life-and-death decisions.
- Killer instinct/"Failure is not an option."
- Cognitive agility in austere environments.
- "Always in season."
- Increased rehabilitation with reduced recovery time.
- Self-reliant.
- Moral and ethical in "grey" circumstances.

The most essential characteristics of the HWS are their mindset and attitude. The ability to make the life-and-death decision to determine mission success and the willingness to never accept defeat set them apart from athletes. Almost any mission for the HWS is high stakes; failure usually means the opposition wins. Failure will result in more injury and suffering for those the HWS tries to protect. The opposition can be anything, from fire and natural disasters to criminals, insurgents, and terrorists. These two traits usually come about based on the austere and life-threatening environments these professionals experience. For any decision made by the HWS, the professional has to consider the third- and fourth-order effects of the decision. This means the HWS needs to analyze the possible outcomes of the possible choices that will be made. This characteristic helps mitigate risks to teammates and civilians and minimize unnecessary damage. However, sometimes these decisions ultimately result in casualties with no way of avoiding the outcome. The HWS always seeks to reduce losses and damage while ensuring mission success. This attribute is something athletes do not experience.

The other concepts to consider during decision-making for the HWS are the **moral and ethical** decisions in "grey" circumstances. In these austere environments, the situation is not always black and white, and the right decision may not be clear. Sometimes it is difficult to determine who is a friend and who is a foe and who is doing right and wrong. The HWS train for this aspect. Some of the training strategies are as follows:

- Vignettes/scenarios
- Stress exercises
  - This is an evolution of vignettes where the HWS performs stressful exercises to induce a high heart rate and fatigue and is then presented with a scenario in which quick decisions and actions must be taken.
- Discussions
- Professional reading

These training strategies are designed to confront the HWS with as many scenarios as possible. Experiences help to elicit the correct split-second decisions while under duress. These types of decisions can determine mission success or loss; these types determine the difference between the lives and deaths of innocents and teammates. Without a doubt, the stakes are very high during these situations; the correct decision must be made, or the consequences could be dire. This is why the HWS spends a significant amount of time training for these scenarios.

The concept of the TA brings about the idea of seasonal training. In sports, teams and athletes have seasons. Each season is based on the 12-month Julian calendar year. A traditional season contains a preseason/base phase, an in-season/competition phase, and a postseason/recovery [1, 22]. During preseason, athletes focus on general strength training, endurance, and recovery [23, 24]. However, this is different for the HWS. These professionals are always in a "competition season." Police officers and first responders do not have off-seasons. These departments rarely have the manpower to rotate officers to have a period off the line of duty to train and increase physical fitness.

> For example, many US firefighter stations have one or two trucks. Since one truck holds about four to six personnel, a station may only have up to ten personnel on duty at one time (the extra personnel in the number includes leadership and administrative personnel). While one crew is on duty, the other crew is off duty. Many firefighters work 24 h on with 48 h off, 10- to 12-h shifts 3 days in a row, or a version thereof. The limited number of personnel assigned to a station, with almost all personnel on or off duty, leaves no room for "base training." The training occurs while on duty, waiting for a call to happen.

The HWS must maintain peak physical and mental conditions while always being ready. This schedule is taxing on these professionals. The lack of downtime can increase injury rates and psychological issues, decreasing combat regeneration and readiness.

The second subtle characteristic difference is "Cognitive agility in highly stressful situations." Cognitive agility is "the capacity of an individual to quickly move back and forth between openness and focus" [25]. The HWS "work in rapidly evolving, complex environments embedded with multiple high-risk factors" [25]. The HWS must be hyperfocused and have cognitive flexibility to succeed in these conditions. Tactical professionals, the police, and first responders need to make split-second decisions that are **morally, ethically, and socially** accepted. However, the HWS must be able to turn it off and reintegrate into daily life. For these professionals, this means going home every night after being in harm's way, coming out of the hyperfocused state, and moving into a state of emotional flexibility and awareness (i.e., openness) [25]. This aspect is much different from the experience of athletes.

Athletes must focus highly during the competition, which comes from cognitive ability (see Table 2.1) [26]. Cognitive ability is "a general mental capability involving reasoning, problem-solving, planning, abstract thinking, complex idea comprehension, and learning from experience" [27]. These abilities are typically seen as relatively stable over time. They are influenced by a combination of genetic factors and early-life experiences and are often thought of as the raw horsepower of the brain. An individual's cognitive abilities can strongly influence their academic and professional achievement.

The HWS must have cognitive ability and agility to be successful. While these two facets of cognition are distinct, they are not entirely separate. A certain level of cognitive ability can provide the foundation for cognitive agility. For instance, a solid ability to reason or problem-solve can aid in adapting to new scenarios. This aspect is necessary since the HWS operates in high-stress and ever-changing 360-degree environments. However, cognitive agility also relies on other factors, such as emotional intelligence, open-mindedness, and resilience, which are not typically included in measures of cognitive ability [25].

This aspect of always "in-season" or "competition phase" and the quantity of available personnel affect the HWS's ability to recover from injury. It is not uncommon for the HWS to not have enough time to recover from injuries sustained in the line of duty [28, 29]. This means the HWS may develop a chronic injury that could be fixed with proper medical attention and recovery time. Returning to the example of CPT McClain, it took him 53 days to learn the prognosis of his ACL injury, whereas an athlete would have surgery within 2 weeks of the injury. HWS professionals typically do not have the time or feel they do not have the time to recover. This aspect means HWS professionals continue to perform injury rehabilitation while performing regular duties. These professionals may not be given the proper time to recover, whether by leadership or themselves, negating the rehabilitation process. The lack of access to medical care and the lack of higher personnel numbers are not usually experienced by elite and professional athletes.

The final aspect of addressing is the self-reliance trait of the HWS. Up to this point, the HWS is typically a part of a team like that of an athlete. However, the HWS must be self-reliant. These professionals tend to find themselves in situations where they are on their own. Many police units will only have one professional in a team. In the military, small numbers aid in stealth and surprise. It is not uncommon for the US Military to work in small teams or individually to gain an advantage over the enemy. Self-reliance is something that is only sometimes encouraged in sports. In team sports, the team is more important than the individual [30]. The HWS must be able to shift integrating into a team and self-reliance seamlessly. This is something other than what is typically encouraged in athletics.

The Definition of HWS

The primary focus of this chapter is to lay out the differences between the HWS and athletes. The focus has been on athletes because the term TA is centered on the athlete paradigm. Furthermore, several phases are highlighted in bold throughout this chapter. The bold phrases are the key terms used to define the HWS. Based on the evidence discussed in this chapter, the HWS can be defined as:

> A human being comprised of a complex set of physiological, cognitive, and emotional systems that create a cohesively functioning tool that is professionally trained, maintained, and optimized; capable of learning and adapting; and using various equipment and knowledge for constructive, destructive, and healing purposes, always bounded by societal moral, ethical, and legal standards.

Conclusion

In conclusion, the human weapon system is a complex and intricate system that is as fascinating as it is powerful. This chapter explored why the concept of "tactical athlete" cannot be applied to tactical, law enforcement and first responder professions.

When trained and optimized, the human body can be a highly effective tool for defense or attack, but this is not the only facet of the human weapon system. This book discusses how the human mind, with its strategic planning, decision-making abilities, and emotional resilience, is just as much a part of this system as the body.

However, a salient point not to be overlooked is the importance of ethics and moral judgment within this context. Our understanding of the human weapon system is intended not to glorify violence but rather to appreciate the human potential and its capabilities. It serves as a reminder of our responsibilities in using these capabilities wisely and ethically, for self-defense, for the protection of others, and for maintaining peace.

Moving forward, we must continue exploring, understanding, and responsibly utilizing the human weapon system, recognizing it as one of the most intricate and evolved aspects of human biology and psychology. A well-interlaced human optimization program will help the HWS reach their maximal potential.

References

1. Alvar B, Sell K, Deuster P. Essentials of tactical strength and conditioning [Online]. 1st ed. Human Kinetics. https://www.nsca.com/store/product-detail/INV/9781450457309/9781450457309
2. Sacco Sez: How the 1997 Broncos broke through in Super Bowl XXXII [Online]. [date unknown]. https://www.denverbroncos.com/news/sacco-sez-how-the-1997-broncos-broke-through-in-super-bowl-xxxii-20331380. 5 Mar 2023.
3. David Wilson CJ, Usa M, David Crawford CA. Preoperative wait time for orthopedic surgeries at a military medical center [Online]. https://academic.oup.com/milmed/article/177/6/740/4283724
4. Leiba A, Weiss Y, Carroll J, Benedek P, Bar-dayan Y. Waiting time is a major predictor of patient satisfaction in a primary military clinic [online]. Mil Med. 2002;167:842. https://academic.oup.com/milmed/article/167/10/842/4819662
5. Emily McDermott C, Joshua Henley SS, Meghan Joyce M, Aden J, Nuelle JAV. Improving care by decreasing wait times: a military health system quality improvement project https://doi.org/10.1093/milmed/usac254.

6. Zain Mehdi. MRI wait times must be shortened to improve health care, cornell researchers say [Online]. The Cornell Daily Sun [date unknown]. https://cornellsun.com/2018/02/08/mri-wait-times-must-be-shortened-to-improve-health-care-cornell-researchers-say/. 25 Feb 2023.
7. Feyman Y, Legler A, Griffith KN. Appointment wait time data for primary & specialty care in veterans health administration facilities vs. community medical centers. Data Brief. 2021;36 https://doi.org/10.1016/j.dib.2021.107134.
8. Taylor M, Hopman W, Yach J. Length of stay, wait time to surgery and 30-day mortality for patients with hip fractures after the opening of a dedicated orthopedic weekend trauma room. Can J Surg. 2016;59:337–41. https://doi.org/10.1503/cjs.017014.
9. Arliani GG, Pereira VL, Leão RG, Lara PS, Ejnisman B, Cohen M. Tratamento das lesões do ligamento cruzado anterior em jogadores profissionais de futebol por cirurgiões ortopedistas. Rev Bras Ortop (Sao Paulo). 2019;54:703–8. https://doi.org/10.1055/s-0039-1697017.
10. How long is recovery time from an ACL tear? UCHealth Today [Online]. [date unknown]. https://www.uchealth.org/today/acl-tears-how-long-does-it-take-to-recover-and-return-to-sports/. 5 Mar 2023.
11. RAND Health. Rand health quarterly. [Online]. https://www.rand.org/pubs/periodicals/health-quarterly/issues/v5/n3/12.html. 3 Mar 2023.
12. Gibbons SW, Hickling EJ, Watts DD. Combat stressors and post-traumatic stress in deployed military healthcare professionals: an integrative review. J Adv Nurs. 2012;68:3–21.
13. Campbell DJ, Nobel OBY. Occupational stressors in military service: a review and framework. Mil Psychol. 2009;21 https://doi.org/10.1080/08995600903249149.
14. Briggs EC, Fairbank JA, Tunno AM, Lee RC, Corry NH, Pflieger JC, Stander VA, Murphy RA. Military life stressors, family communication and satisfaction: associations with children's psychosocial outcomes. J Child Adolesc Trauma. 2020;13 https://doi.org/10.1007/s40653-019-00259-z.
15. Thelwell RC, Weston NJV, Greenlees IA, Hutchings NV. Stressors in elite sport: a coach perspective. J Sports Sci. 2008;26 https://doi.org/10.1080/02640410801885933.
16. Hanton S, Fletcher D, Coughlan G. Stress in elite sport performers: a comparative study of competitive and organizational stressors. J Sports Sci. 2005;23 https://doi.org/10.1080/02640410500131480.
17. Egan KP. Supporting mental health and well-being among student-athletes. Clin Sports Med. 2019;38:537–44.
18. Adler AB, McGurk D, Stetz MC, Bliese PD. Military occupational stressors in garrison, training and deployed environments. NIOSH/APA Symp. 2003;298:1–17.
19. Kaikkonen NM, Laukkala T. International military operations and mental health – a review. Nord J Psychiatry. 2016;70:10–5.
20. Rice SM, Purcell R, de Silva S, Mawren D, McGorry PD, Parker AG. The mental health of elite athletes: a narrative systematic review. Sports Med. 2016;46:1333–53.
21. Mind, body and sport: the psychiatrist perspective – NCAA.org [Online]. [date unknown]. https://www.ncaa.org/sports/2014/11/3/mind-body-and-sport-the-psychiatrist-perspective.aspx. 3 Mar 2023.
22. Bompa TO. Periodization: theory and methodology of training. 4th ed. Champaign: Human Kinetics; 2009.
23. Bompa T, Buzzichelli C. Periodization: theory and methodology of training. 6th ed. Champaign: Human Kinetics; 2018.
24. Haff G, Triplett NT. Essentials of strength training and conditioning. Champaign: Human Kinetics; 2014.
25. Ross J, Miller L, Deuster PA. Cognitive agility as a factor in human performance optimization. J Spec Oper Med. 2018;18:86–91.
26. Staiano W, Merlini M, Romagnoli M, Kirk U, Ring C, Marcora S. Brain endurance training improves physical, cognitive, and multitasking performance in professional football players. Int J Sports Physiol Perform. 2022;17:1732–40. https://doi.org/10.1123/ijspp.2022-0144.

27. Gottfredson LS. Why g matters: the complexity of everyday life. Intelligence. 1997;24:79–132. https://doi.org/10.1016/S0160-2896(97)90014-3.
28. Hando BR, Bryant J, Pav V, Haydu L, Hogan K, Mata J, Butler C. Musculoskeletal injuries in US Air Force Tactical Air Control Party trainees: an 11-year longitudinal retrospective cohort study and presentation of a musculoskeletal injury classification matrix. https://doi.org/10.1136/military-2023-002417
29. Abbott A, Wang C, Stamm M, Mulcahey MK. Part II: risk factors for stress fractures in female military recruits. Mil Med. 2023;188:93–9.
30. Kim SS, Baek WY, Byon KK, Ju SB. Creating shared value and fan loyalty in the Korean professional volleyball team. Sustainability (Switzerland). 2020;12 https://doi.org/10.3390/su12187625.

Chapter 3
What Is Human Performance Optimization?

Introduction

In the previous chapter, the argument was made on changing the prism through which first responders and tactical professionals are viewed and analyzed. The human weapon system (HWS) has an entirely different set of skills and needs than athletes. In this chapter, the argument will be made on addressing these needs. Much of the background information found in the chapter will be focused on the Department of Defense (DoD). This aspect is due to few details published on human performance programs for first responder professions.

Human Performance (HP)

The concept of human performance (HP) is a concept that has been introduced previously. For example, elite and professional sports teams have used this training theory for decades. However, the idea of HP is relatively new to the HWS career fields. The first HP programs were introduced for the DoD around 2005 [1]. Later, in 2009, the DoD created the Human Performance Resource Center (HPRC), with the task of collecting and publishing "fitness, well, and performance" information for the DoD [2]. Throughout the upcoming years, the DoD observed significant increases in self-harm and suicide rates with drops in medical readiness due to the high operational tempo of the Iraq and Afghanistan conflicts. So in 2011, the DoD created Total Force Fitness (TFF) to "enhance health and optimize performance across eight domains: physical, environmental, medical and dental, nutritional, spiritual, psychological, behavioral, and social fitness" [1, 3]. Since then, HP programs in the DoD have morphed into several different models, which will be discussed later in this chapter.

C. Myers, *The Human Weapon System*,
https://doi.org/10.1007/978-3-031-45061-7_3

With the evolution and application of HP programs, HP has been relatively undefined. Several pieces of literature define the "factors" or "structure" of HP. Some of the existing definitions can be found as follows:

(a) "A series of behaviors executed to accomplish specific results (performance = behavior + results)" [4]
(b) "A New Training paradigm that calibrates to the unique physiological needs of individual athletes, and the differing operation demands of any team or organization" [5].

The highly evident aspect is that many HP definitions must focus more on a particular profession or population. Therefore, the definitions are very general. So an HP definition for the HWS needs to be developed. In the previous chapter, Table 2.1 laid out the essential characteristics of the HWS. The HWS definition and characteristics need to be included in the HWS version of the HP definition. With these characteristics in mind, the HWS HP definition is as follows:

> HP is a training and maintenance paradigm utilizing the pillars or human factors to improve human physiological and cognitive responses and adaptations to austere environmental factors, improve regeneration and resiliency, and improve medical care from a reactive to a preventative state.

The traditional HP model (Fig. 3.1) uses pillars or "human factors" [6]. These pillars can be any number and type of factors deemed necessary for the HP program to achieve its focus. The focus of the HP program is essential; it is why an organization should have this type of program.

One example is an HWS organization that wants to improve muscular strength, endurance, and power and improve combat regeneration; the HP model must address the factors that create this type of performance focus. Five factors were identified during the needs analysis process (which will be discussed in further detail in Chap. 9) [7]. The five factors are as follows: fitness, recovery, nutrition, cognitive, and medical. The HP team identified these pillars necessary to achieve the objective of their program (Fig. 3.2).

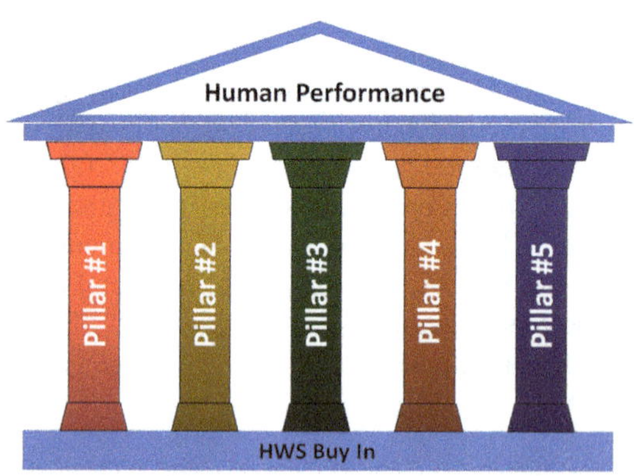

Fig. 3.1 The traditional human performance model. This model utilizes pillars (aka human factors) to describe the areas of training focus to achieve human performance

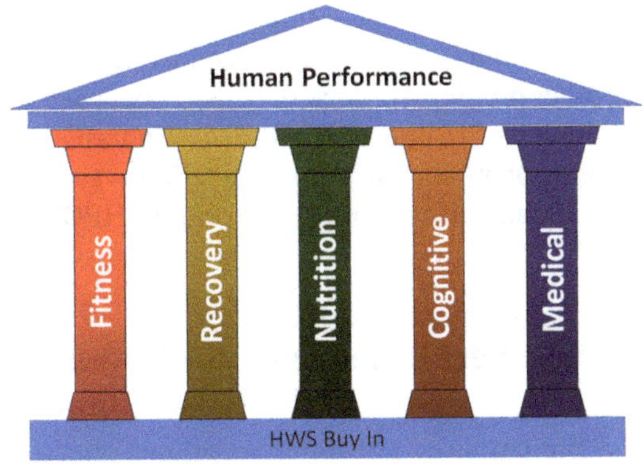

Fig. 3.2 A traditional 5-pillar HP program improving muscular strength, power, and recovery

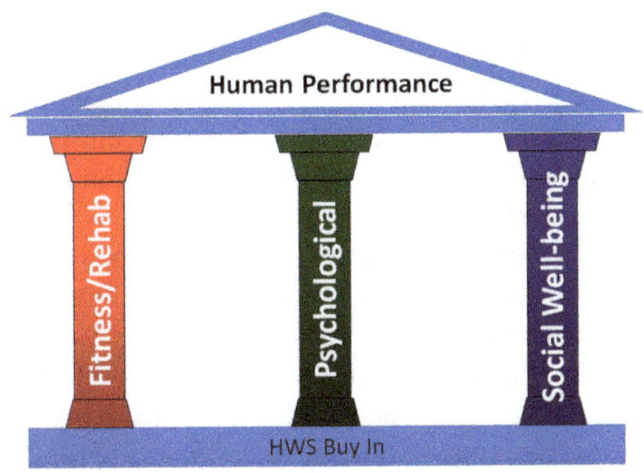

Fig. 3.3 A traditional three-pillar HP program focusing on mental health

A second example is a program focusing on the mental health of the HWS. Through the needs analysis process, the HWS HP team identified three pillars to address. These pillars are physical fitness, rehabilitation, and psychological and social well-being (Fig. 3.3).

As these two HP program examples demonstrate, an HP program can have any number and type of pillars. The number and type of pillar depend on the HP program's focus. As stated earlier, the type and quantity of pillars are identified during the needs analysis process (Chap. 9). Differences aside, all successful HP programs have one item in common; this foundation is HWS buy-in.

The HWS buy-in is the foundation because any HP program needs to work. The HWS will not embrace the program if they feel it is not for them. Therefore, a successful HP program must communicate why and how it will improve the performance of the HWS. This concept will be discussed in further detail in Chap. 9.

This HP pillar model has its merits but has one significant limitation. The definition of a pillar states that limitation. A pillar is a "firm, upright support or a column or shaft standing alone" [8]. This definition of a pillar states that each human factor is separate; these types of programs inherently have no integration. This means that each pillar does not interact with the other pillar. HP professionals in these types of programs tend not to communicate with each other. These professionals rely on the HWS to communicate what is happening in the other pillars or to completely ignore the other pillars. This lack of synergy between the pillars prevents the program from reaching its full potential and focus. In turn, the customer, the HWS, does not reach its full potential. Since this limitation was identified, the HP model has morphed into human performance optimization (HPO).

Human Performance Optimization (HPO)

The limitation of HP pillar programs can be rectified by reevaluating pillars as domains and integrating the domains. A domain is "a sphere of knowledge, influence, or activity" [3, 9]. Changing the fundamental structure of the HP paradigm from a pillar to a domain changes the entire HP approach. A domain is one part of a whole program. By itself, it cannot achieve the end state. The domain must work with the other domains to achieve the ultimate goal. However, only changing the pillars to domains does not fix the traditional HP program limitations. Integration is the answer.

Interlacing the different domains means each domain must work hand in hand to achieve the program's objective. When one domain does not work with the others, the program does not assist the HWS in reaching and maintaining their full potential. In addition, an interlaced approach allows you to add and change out cogs to make the program adaptable and better. This aspect is how an HP program is optimized, making it a human performance optimization (HPO) program. Based on the changes in domains and integration, the definition of an HP program can evolve into an HPO program as follows (Fig. 3.4):

> HPO is a multifactorial, holistic, **interlaced** training and maintenance paradigm to **improve** human physiological and cognitive responses and adaptations to austere environmental factors, regeneration and resiliency, and medical care from a reactive to a preventative state.

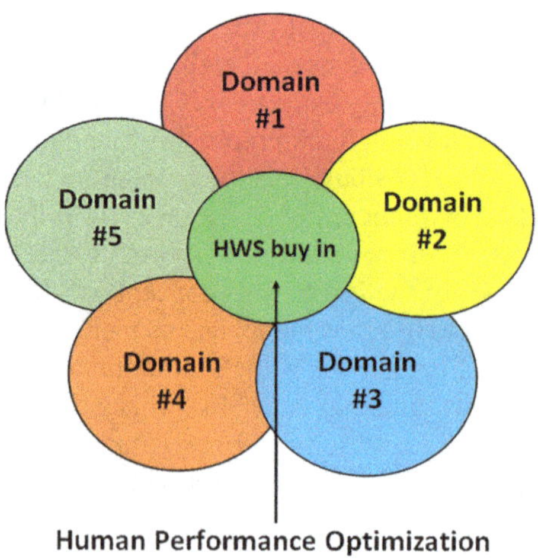

Fig. 3.4 Basic human performance optimization model

Similarly to the pillar HP model, HWS buy-in is essential to the program's success. In the case of the pillar model, buy-in was the foundation. However, the domain model is different; buy-in is the binding force that keeps the intersections of all the domains together and functioning cohesively [9]. Without the HWS buy-in, the best HPO programs will not be successful in their charter.

The domains are determined, similarly to pillars, through the needs analysis process (Chap. 9). The primary domains seen in HPO programs are as follows:

1. Musculoskeletal Domain
2. Nutritional Domain
3. Cognitive Domain
4. Psychological Domain
5. Spiritual Domain
6. Medical Domain

These listed domains are not all exhaustive; however, these are the main six domains (in various combinations) seen in most HPO programs. Other possible domains are Education and Life-Cycle Management; the needs analysis determines the domains used by the HPO program. Typically, HPO domain models have more domains than HP pillar models. An example of this difference is seen in the TFF document to the DoD produced in 2011. The DoD recognized that human performance was integral to improving the health of the HWS and required a more holistic approach than what was typically used. In 2013, the first HPO program was implemented by the DoD.

The Preservation of the Force and Family (POTFF) was introduced by the Special Operations Command (SOCOM). The POTFF program was created for the following purpose:

> The mission of the Preservation of the Force and Family (POTFF) is to optimize and sustain Special Operations Forces (SOF) mission readiness longevity, and performance through integrated and holistic human performance programs designed to strengthen the Force and Family (SOCOM POTFF).

The end state of this program is as follows:

> The end-state of the POTFF is a resilient SOF enterprise with optimized performance across all domains enabling improved readiness to the best meet the operational demands for the duration of a SOF career (SOCOM POTFF).

As shown in Fig. 3.5, the POTFF program is a four-domain model of human performance (physical), psychological performance (mind), social performance (social and family), and spiritual performance (spirit) (SOCOM POTFF) that focuses on resiliency. These domains are based on the SOF community's needs and the program's mission. The interlinking component of this model is the Operator and family; this is the buy-in element of the program [10].

Other HPO programs exist within the HWS professions, such as True North (Air Force), Holistic Health and Fitness (H2F) (Army), and Integrated Occupational Support (IOS) (Air Force). The HPO approach is being embraced by North Atlantic Treaty Organization (NATO) organizations as well. For example, the Multinational Capability Development Campaign published a Human Performance Optimization and Enhancement report in March 2021 [1]. This report outlined the different domains and how to implement this type of HPO program across various organizations. The basic model structure of these listed programs is similar; the differences occur in the focus of each program. This aspect of HPO programs will be discussed in the next section of this book.

A New Direction for HPO

Up until this point, the focus definition of HPO has been "to **improve** human physiological and cognitive responses and adaptations to austere environmental factors, regeneration, and resiliency." However, the second part of the HPO definition says, "medical care from a reactive to a preventative state." Looking at HPO as a new

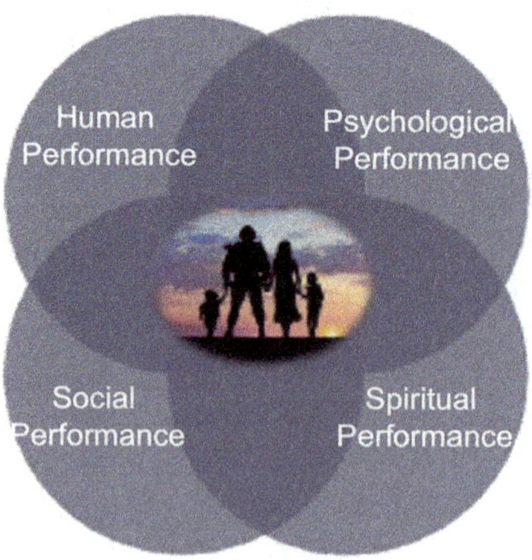

Fig. 3.5 Preservation of force and family model (SOCOM POTFF)

form of medical care is a new application for this type of training paradigm in the HWS professions [1, 3, 11, 12].

Research has established that the main HPO domains have a synergistic play on each other. Each domain works in tandem to improve an individual's health [1, 3, 5, 6, 11, 13]. Furthermore, HPO programs are now seen as a preventative medical model compared to the current reactive medical model [1, 13]. The current state of most medical programs is reactive. This means you seek medical attention after you get injured or sick. However, what if the current medical treatment was changed?

Imagine medical treatment as a continuum, as shown in Fig. 3.6. On the left side of the continuum is preventative medicine. Preventative medicine aims to prevent injury or disease before it begins [14]. This type of treatment includes measures that look to improve all the domains of the HWS to mitigate injury and illness. Based on the discussion of the basis of HPO models earlier in this chapter, HPO programs inherently become preventative medicine tools by operating to make the HWS work at optimal levels.

Reactive medicine is on the other side of the continuum, as shown in Fig. 3.6. This is the current state of medicine. Referring to Chap. 2, CPT McClane received medical attention after his injury. This example demonstrates reactive treatment. Now, all injuries cannot be prevented. However, preventative medicine can help reduce the occurrence and severity of injuries.

CPT McClane's example was used in Chap. 2 due to the ACL injury. ACL injuries have a high occurrence rate in HWS professions [15]. However, evidence-based preventative medicine strategies can significantly reduce ACL injury rates [15–17]. These strategies include neuromuscular priming methods, task-specific speed and agility drills, dynamic stretching, and balancing routines [15]. These preventative strategies help improve joint range of motion and skeletal muscle strength and power. Putting these elements together helps the knee to absorb more rotational force without rupturing the ligament. Ultimately, these strategies can differ between a fully ruptured ACL (recovery time of 18–24 months) and a partially ruptured ACL (recovery time of 3–6 months) (Fig. 3.7).

When the HPO model is applied to the healthcare continuum, certain domains are more active than others. This is highly dependent on where the HWS is located on the continuum. For example, in the Musculoskeletal Domain, if the HWS is uninjured, HWS will spend most of the time interacting with the physical performance staff. However, this does not mean the HWS will not interact with the rehabilitation team. The rehabilitation team does spend time with healthy HWS professionals who are uninjured. This concept will be discussed in more detail in Part II. A well-designed and implemented HPO program can be a preventative and

Fig. 3.6 Healthcare continuum

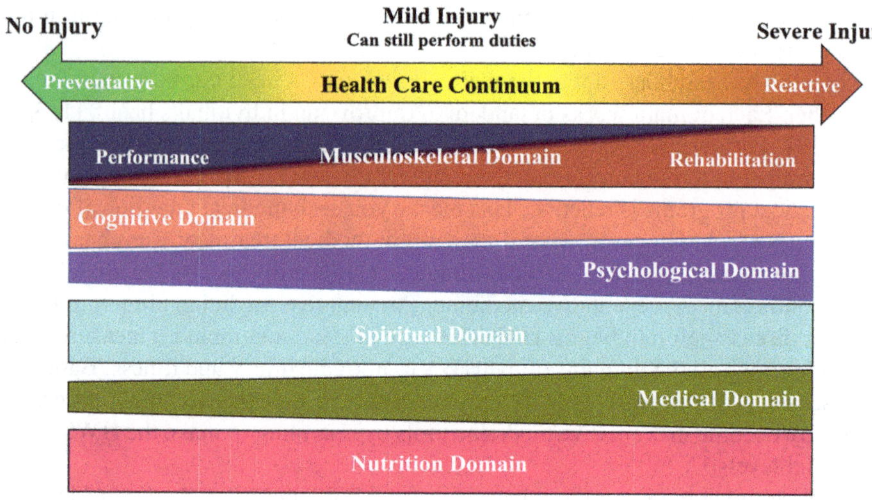

Fig. 3.7 The HPO Model Continuum with regard to the Health Care Continuum. This continuum contains the primary six domains seen within most HPO programs. The order of these domains complements each other

reactive medicine tool. It can reduce the number and severity of injuries and reduce rehabilitation time to improve and maintain readiness and mission success.

Conclusion

HPO is a multifactorial, holistic, interlaced training and maintenance paradigm to improve human physiological and cognitive responses and adaptations to austere environmental factors, regeneration and resiliency, and medical care from a reactive to a preventative state. It transcends the simple notion of physical fitness or intellectual prowess and embraces a holistic perspective that includes health, mental resilience, recovery, nutrition, and optimizing individual and team behaviors.

HPO is a continuous, dynamic process. It is not merely about reaching a peak but is about consistently enhancing and maintaining our physical and mental performance over time. HPO is about sustainably maximizing human potential and, in doing so, improving quality of life, productivity, and longevity, and moving healthcare from a reactive state to more of a preventative state.

References

1. Haggenmiller C. Human performance optimization and enhancement. Oxford: Oxford University Press; 2021.
2. Alvar B, Sell K, Deuster P. Essentials of tactical strength and conditioning [online]. 1st ed. Human Kinetics https://www.nsca.com/store/product-detail/INV/9781450457309/9781450457309
3. Chairman's Total Force Fitness Framework [Online]. http://www.dtic.mil/cjcs_directives
4. Borman WC. Job behavior, performance, and effectiveness. In: Handbook of industrial and organizational psychology, vol. 2. 2nd ed. Washington, DC: APA; 1991.
5. Human performance training for operational readiness [Online]. [date unknown]. https://www.boozallen.com/expertise/digital-solutions/human-performance.html. 25 Mar 2023.
6. Matthews G, Davies DR, Westerman SJ, Stammers RB. Human performance: cognition, stress and individual differences. Hoboken: Taylor and Francis; 2013.
7. Myers C, Allen H. Triathlon training with power [Online]. 1st ed. Cognella Publishing https://titles.cognella.com/triathlon-training-with-power-9781793532725. 11 Sept 2021.
8. Pillar definition [Online]. [date unknown]. https://www.merriam-webster.com/dictionary/pillar?utm_campaign=sd&utm_medium=serp&utm_source=jsonld. 25 Mar 2023.
9. Domain definition [Online]. [date unknown]. https://www.merriam-webster.com/dictionary/domain. 25 Mar 2023.
10. SOCOM POTFF [Online]. [date unknown]. https://www.socom.mil/POTFF/Pages/default.aspx. 25 Mar 2023.
11. Deuster PA, Grunberg NE, O'Connor FG. An integrated approach for special operations. J Spec Oper Med. 2014;14 https://doi.org/10.55460/hetg-w3e1.
12. Sell TC, Lutz RH, Faherty MS. The warrior model for human performance optimization. Sports Med Arthrosc Rev. 2019;27:99–106. https://doi.org/10.1097/JSA.0000000000000248.
13. Deuster PA, Grunberg NE, O'Connor FG. An integrated approach for special operations. J Spec Oper Med. 2014;14:86–90. https://doi.org/10.55460/hetg-w3e1.
14. Clarke EA. What is preventive medicine? [Online]. Can Fam Physician. 1974; 20: 65–8. http://www.ncbi.nlm.nih.gov/pubmed/20469128. 25 Mar 2023.
15. Herman K, Barton C, Malliaras P, Morrissey D. The effectiveness of neuromuscular warm-up strategies, that require no additional equipment, for preventing lower limb injuries during sports participation: a systematic review. BMC Med. 2012;10:75. https://doi.org/10.1186/1741-7015-10-75.
16. Grimm NL, Jacobs JC, Kim J, Denney BS, Shea KG. Anterior cruciate ligament and knee injury prevention programs for soccer players. Am J Sports Med. 2015;43:2049–56. https://doi.org/10.1177/0363546514556737.
17. Paulson W, Slattengren AH. Effectiveness of ACL injury prevention programs [Online]. Am Fam Physician. 2018;97 [Online]. https://www.aafp.org/pubs/afp/issues/2018/0201/od2.html. 25 Mar 2023.

Part II
The Human Performance Domains in Human Performance Optimization Programs

Chapter 4
The Musculoskeletal Domain

Introduction

The Musculoskeletal (MSK) Domain is primarily emphasized in HPO programs. This domain is designed to improve physical performance and the human weapon system (HWS) physiology. In addition, physical fitness improves cognitive and psychological functions (Chap. 6) and recovery. This chapter will discuss the baseline personnel composition, application, testing, and analytics used within this domain.

The Musculoskeletal (MSK) Domain

The term MSK Domain is relatively new. Traditionally, this domain exists as two separate domains: performance and rehabilitation. The performance domain focuses on improving physical performance through resistance and endurance training. A strength and conditioning (SC) coach leads the performance domain. The rehabilitation domain focuses on the return from injury. A physical therapist (PT) leads the rehabilitation domain. Recently, these two domains have been combined into a single domain termed the MSK domain. The MSK domain focuses on improving physical performance, minimizing injury, and improving recovery times. Before diving into this explanation further, let us discuss the typical composition of the MSK Domain team.

C. Myers, *The Human Weapon System*,
https://doi.org/10.1007/978-3-031-45061-7_4

The MSK Domain Occupation Composition

The MSK Domain will consist of several different occupations. Each occupation assists each focus (physical and rehabilitation) in achieving their end states. Therefore, the MSK team composition may consist as follows.

Performance Team

Exercise Physiologist (EP)

EPs analyze fitness data and interpret the data to develop fitness and exercise programs to improve clients' musculoskeletal performance and overall health. The primary role of an EP is to integrate with Department of Defense (DoD) research teams to understand and apply new training methods and information to help the HWS optimize performance in austere conditions. Furthermore, EPs program performance training and coaching while providing input to the command and operations team to improve HWS performance during mission execution and recovery.

These professionals typically have a bachelor's degree from an accredited institution. Exercise physiology degree programs have cardiovascular physiology, anatomy, nutrition, strength and conditioning, biology, and kinesiology courses and are recognized as allied health professionals. In addition, clinical-oriented exercise physiology programs will include clinical work [1, 2]. Furthermore, exercise physiologists have certifications from certified organizations such as the American College of Sports Medicine (ACSM), the National Strength and Conditioning Association (NSCA), the American Red Cross, etc. Some certifications include Certified Strength and Conditioning Specialist (CSCS), ACSM-Exercise Physiologist, CPR, and basic nutritionist certification.

Important note: EPs are not considered providers. Even though they work with many medical professionals, they do not have the same privileged rights as PTs do. The Defense Health Agency (DHA) defines privileged providers as possessing "appropriate credentials and are granted authorized clinical privileges to diagnose, initiate, alter, or terminate regimens of healthcare with a defined scope of practice" [3]. This means an EP (in the military and the civilian sector) cannot diagnose or treat injuries and diseases. An EP must refer any injured or sick HWS to the correct professional (i.e., PT or medical doctor).

Strength and Conditioning (SC) Coach

A SC coach is an individual who holds a certification from an accredited organization in strength and conditioning. The primary focus of the SC coach is to program performance training and performance coaching. The most notable certifications are the NSCA CSCS and USA Weightlifting Coaching certifications. The CSCS certification is more stringent because it requires individuals to hold a minimum of a bachelor's degree or a terminal degree in physical therapy or chiropractic medicine [4]. To be a SC coach, one does not have to have an academic background or degree in this area unless dictated by a certifying organization.

**Note: SC coaches are not medical personnel like many other MSK Domain team members.

Certified Athletic Trainer (ATC)

This profession is discussed in the Rehabilitation Team section.

Rehabilitation Team

Physical Therapist (PT)

PTs examine the HWS, diagnose and treat injuries, reduce or manage pain, and improve and restore function [5]. PTs are privileged providers, which allows these professionals to diagnose and treat injuries. This aspect of the PT profession is critical to the MSK Domain team since EPs cannot diagnose or treat injuries. This aspect is why EPs, SC coaches, and PTs work together along all aspects of the Healthcare Continuum (Figs. 4.1 and 4.4).

The American Physical Therapy Association (APTA) governs PT certification and licensure. A PT must graduate with a Doctor of Physical Therapy (DPT) from an accredited academic institution and pass the state licensure exam [5]. A DPT program typically contains 80% classroom and labs, with the other 20% dedicated to clinical experiences [5]. Usually, to be a PT in the military, the licensure may come from any state in the United States, the District of Columbia, Puerto Rico, or the US territory.

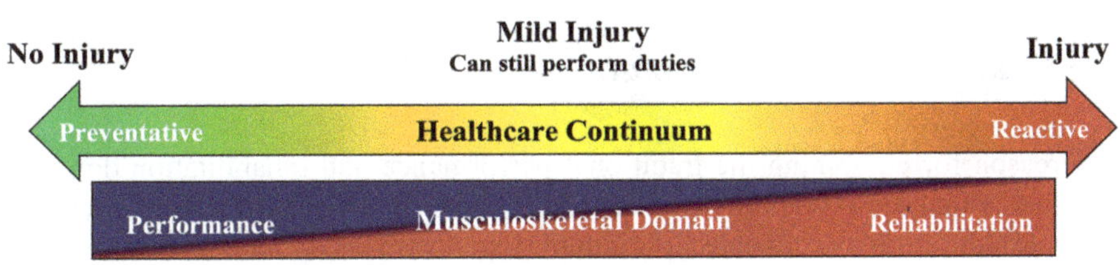

Fig. 4.1 The MSK domain about the Healthcare Continuum

ATC

The primary focus of an ATC is to “keep the HWS going or get HWS back into the mission” [6]. ATC will focus on return-to-duty activities, including working on a range of motion (ROM) and movement deficiencies, applying PT rehabilitation guidance, and working with the EP/SC to maintain or minimize performance degradation. As discussed later in this chapter, the ATC is the bridge between the performance and rehabilitation teams [6].

An ATC is governed by certifying organizations such as the National Athletic Trainer Association (NATA), Board Certification for the Athletic Trainer (BOC), and Commission on Accreditation of Athletic Training Education (CAATE). To become certified, one must have a terminal bachelor’s or master’s degree from an accredited academic institution. In addition, the institution must be accredited by the National Athletic Trainers’ Association (NATA) in athletic training [7]. Many ATCs will hold certifications in SC and nutrition as part of their training. This aspect allows ATCs to be a part of the physical *and* rehabilitation teams. This aspect of ATCs will be discussed in more detail in the “Application of the MSK Domain” section.

In the civilian realm, ATCs are recognized as allied health professionals [8]. However, the Defense Health Agency does not recognize ATCs as allied health professionals [9]. This aspect within the US military limits the scope of practice for ATCs [6].

Physical Therapy Assistant (PTA)

The focus of PTAs is to assist PTs in executing rehabilitation plans. As with PTs, PTA certification is governed by the APTA. A PTA must graduate from the 2-year-long Commission on Accreditation in Physical Therapy Education program and pass the PTA nationally recognized licensure exam. PTAs assist PTs with treating clients but work under the PT’s supervision. The PTA does not have the autonomy of a PT or an ATC. To have a PTA as part of the rehabilitation team, the team must have a PT [10].

Application of the MSK Domain

The reason for combining the traditional performance and rehabilitation domains into one is that these domains are part of the same focus. However, they focus on separate sides of the Healthcare Continuum (Fig. 4.1). This domain is termed “musculoskeletal” because performance and rehabilitation specialists concentrate on the muscles and bones of the human body. To put this into context, refer to Fig. 4.1.

In this figure, the MSK Domain has two sides: physical and rehabilitation. With this overlayed on the Healthcare Continuum, the physical side of the MSK Domain plays more of a role in the preventative side of the continuum. As one moves to the right of the continuum closer to the reactive side, less of the physical side is utilized while more of the rehabilitation side is used. This schematic means the more an HWS is injured, the less interaction the HWS will have with the physical performance team and more interaction with the rehabilitation team. Only when the HWS is injury free that the HWS can focus on being in peak physical condition.

Another aspect of the MSK Domain that can be gleaned from Fig. 4.1 is how the team interacts with each other. As stated earlier, the physical and rehabilitation units are two sides of one coin. These teams work together (Fig. 4.3). A good MSK team meets regularly to discuss and often trains clients. This concept means that when an HWS is not injured, the HWS will interact with the rehabilitation team, albeit smaller. The inverse is confirmed with an injured HWS.

As mentioned in the previous section, an ATC can work with both teams within the MSK Domain. This aspect is due to the ATC academic curriculum and certifications. Many ATCs will have CSCS and nutritionist certifications. Combining these certifications with an ATC academic curriculum, an ATC brings a wealth of knowledge to the MSK Domain.

As shown in Fig. 4.2, an ATC can provide a wealth of expertise within the MSK Domain. Unlike physical/performance and rehabilitation/recovery teams, ATCs do not have to limit their interactions with the HWS. Figures 4.1 and 4.2 depict the preventative and reactive sides of the Healthcare Continuum. When injured, the HWS will spend most interaction time with the rehabilitation team and less with the performance team. The same is true when the HWS works on peak fitness or preventative. Most of the HWS time is spent with the performance team and less with the rehabilitation team (Fig. 4.4).

The flexibility of the ATC allows this professional to migrate between both groups (Fig. 4.4). An HWS can interact with an ATC throughout the Healthcare Continuum.

As shown in Fig. 4.4, an HWS interaction with this MSK team changes depending on the HWS' status. On the prevention side (i.e., no injuries) of the Healthcare Continuum, the HWS interacts more with the performance team (EP/SC + ATC) than with the rehabilitation team (PT & PTA). The HWS interaction with the performance and rehabilitation teams begins to change as injury status begins to change. Moving toward the center of the Healthcare Continuum (mild injury or rehabilitating a chronic injury), the HWS interaction between the performance and

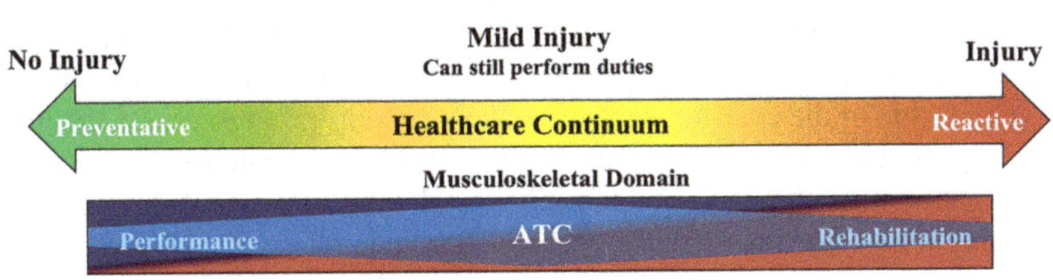

Fig. 4.2 The ATC profession's relationship to the MSK domain teams. (This description of the ATC is adapted from the presentation given by Ms. Patricia Long in February 2023)

Fig. 4.3 Synergistic approach to the entire MSK domain team

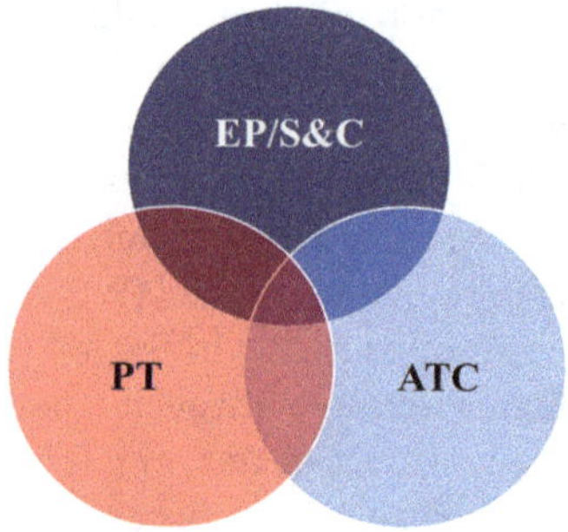

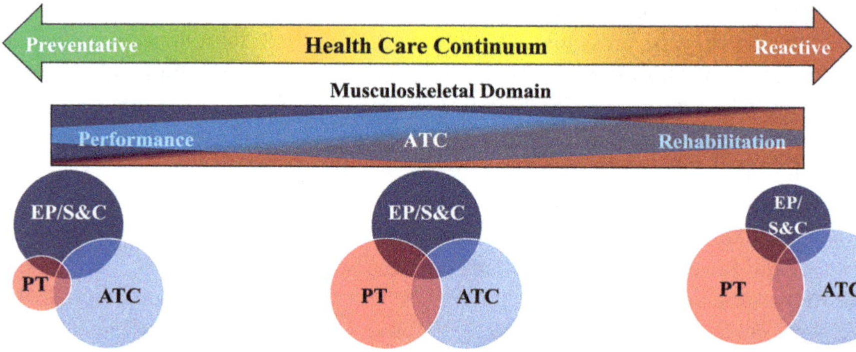

Fig. 4.4 MSK domain team synergy concerning the Healthcare Continuum. (Note that this diagram is modified from the original MSK domain within the Healthcare Continuum developed by Dr. Myers and Dr. Manon and the ATC presentation authored by Ms. Patricia Long)

rehabilitation teams tends to be equal. This equal-part interaction focuses on assisting in rehabilitating the injury while maintaining the HWS's current fitness level. The HWS can still perform his duties, possibly with some limitations, with a mild injury. This is why maintaining fitness while rehabilitating the injury is critical. When an injury prevents the HWS from performing duties (rehabilitation side of the Healthcare Continuum, Fig. 4.4), the primary interaction is with the rehabilitation team and the ATC. This interaction focuses on recovering from the injury and returning the HWS to performing his duties. The performance team assists with performance programming based on the guidance of the rehabilitation team.

The one consistent aspect throughout this process is interaction with an ATC. An ATC can help bridge the transition of an injured HWS from the reactive side to the preventative side of the continuum as the HWS recovery progresses. On the prevention side of the Healthcare Continuum (Fig. 4.4), the ATC can assist with the day-to-day maintenance of a range of motion (ROM) and movement decrements [6]. This interaction of the ATC with the performance team allows the EP/&SC coach to focus on training programming and coaching. The ATC duties shift when working with an HWS who is mildly injured or focusing on a chronic injury. The ATC focuses on adapting the performance team's training to meet the mildly injured

HWS needs while working with the PT with injury treatment [6]. This is an aspect of the duties of the ATC that stands out. The ATC speaks the language of performance and rehabilitation. The ATC is an essential bridge for the HWS to maintain performance or minimize performance decrements while in a mildly injured state. Finally, on the far right of the Healthcare Continuum (Fig. 4.4), the HWS's primary interaction is with the rehabilitation team, with less interaction with the performance team. The PT evaluates and prescribes injury rehabilitation protocols, while the ATC works with both teams (performance and rehabilitation). The performance team works with the rehabilitation team to program training and coaching that compliments the rehabilitation team's rehabilitation protocol for the HWS.

The HWS's interaction with MSK teams in the MSK Domain changes based on the injury's status. By having a synergy between the performance and rehabilitation teams (Fig. 4.3), the teams work together to ensure the injuries are mitigated, recovery times are optimal, and performance is maintained and optimized.

Specific Applications and Performance Metrics of the MSK Domain

Performance Team

Applications

Some of the areas on which the performance team can focus their efforts are as follows:

1. ROM and flexibility
2. Muscular strength
3. Muscular power
4. Muscular endurance
5. Aerobic capacity
6. Anaerobic capacity
7. Speed
8. Agility
9. Neurocognitive agility
10. Anthropometrics

The specificity of focus areas is determined by what is gleaned from the needs analysis (Chap. 9), mission requirements, and command team priorities. Then the prioritized aspects are planned into a comprehensive, periodized training program. This book will not go into this process because other excellent resources and textbooks (such as Bompa's *Periodization* and the NSCA's *Essentials of Strength and Conditioning*) go into detail.

Performance Metrics

As part of any training program, performance tests must be conducted to determine if the training program works for the HWS. Some of the most utilized, scientifically validated performance tests are listed below. Further details on each test can be found in the cited references.

1. ROM and flexibility
   - (a) Sit and reach test [11]
   - (b) Overhead squat test [11]
2. Strength
   - (a) One-repetition max (1RM) [11, 12]
   - (b) Three-repetition max (3RM) [11, 12]
   - (c) Velocity based [13, 14]
3. Power
   - (a) Upper body
     - (i) Velocity based [13, 14]
     - (ii) Medicine ball chest pass [11, 15]
     - (iii) Bench press throw [11, 15]
     - (iv) Upper body Wingate test [11, 15]
   - (b) Lower body
     - (i) Countermovement jump (CMJ) – double- and single-leg variations [11, 15]
       1. This test can be performed with or without a force plate. Force plates will be discussed more in Chap. 12.
     - (ii) Squat jump (SJ) [11, 15]
     - (iii) Broad jump [11, 15]
     - (iv) Lower body Wingate [11, 15]
4. Muscular endurance
   - (a) Upper body
     - (i) Push-up [11, 16]
     - (ii) YMCA bench press test [11]
     - (iii) Front plank test [16]
   - (b) Lower body
     - (i) Wall sit – double- and single-leg variations
     - (ii) 30-second hurdle jump test

5. Aerobic capacity
   (a) Threshold tests
      (i) Run tests
         1. 5-km run [17, 18]
         2. 1.5-mile run [11, 16, 19]
         3. Yo-Yo intermittent recovery test [11, 19]
      (ii) Swim tests
         1. 1500-m swim test [17, 18]
         2. 500-m swim test [19]
      (iii) Threshold bike tests
         1. 20-min bike tests [17, 18]
         2. 10-mile time trial [19]
      (iv) Rowing threshold tests
         1. 20-min rowing test [19]
6. Anaerobic capacity
   (a) 300-yard shuttle run [11]
7. Speed
   (a) Straight-line sprint tests [11, 20]
8. Agility
   (a) Proagility test (aka 5-10-5) [11, 21]
   (b) T-test [11, 21]
   (c) Y-test [11, 21]
   (d) Hexagon test [11]
   (e) Three-cone test [21]
   (f) Reactive agility test [21]
9. Neurocognitive agility
   (a) Reactive drill tests [21]
   (b) Hexagon test [11]
10. Anthropometrics
   (a) DEXA scan
   (b) Three- and seven-site skinfolds [11]
   (c) BodPod testing

Rehabilitation Team

Applications

The priority of the rehabilitation team is to correctly diagnose the HWS injury and create a rehabilitation plan for return to duty. To perform this, PTs have a systematic approach to diagnosis. One of the ways is through a three-step process, which is as follows [22]:

- Question 1: What are the patient's symptoms or injuries? How serious is the injury? What is the pain level?
- Question 2: Where is the pain originating?
- Question 3: What occurred to cause the pain?

In trying to answer these three questions, the PT has a battery of tests to determine the location and severity of the injury [22].

Diagnostic Tests

Each PT will have a systematic way of obtaining a differential diagnosis for an injury. However, the most used diagnostic tests are as follows [22]:

1. Obtaining client history
   (a) How did the injury occur?
   (b) Where is the injury? Is there any pain, swelling, or instability?
   (c) What is the joint/limb's overall function?
   (d) Is there any other history of injuries?
   (e) What is the patient's overall health?
   (f) Is there any possibility of overuse injuries?
   (g) Is the injury acute or chronic?
   (h) Are there other factors, such as nutritional, pharmaceutical, psychosocial, etc.?
2. Physical assessment
   (a) Assessment of the joints and musculoskeletal tissues associated with the pain.
   (b) Assessment of mechanosensitivity.
   (c) Assessment of the spine.
   (d) Assess of biomechanics.
      (i) Walking.
      (ii) Running.

(e) Perform a functional movement assessment.

  (i) Squat or overhead squat test.
  (ii) Any movement related to the HWS movement.

(f) ROM testing.
(g) Palpitations.
(h) Assessment of the ligaments.
(i) Strength testing.
(j) Neural mechanosensitivity assessments.

  (i) Slump test.
  (ii) Straight leg raise.
  (iii) Neural Thomas test.
  (iv) Upper limb neurodynamic test.

(k) Movement technique.
(l) Equipment.

3. Imagery assessment

  (a) X-rays.
  (b) Magnetic resonance imaging (MRI)
  (c) Ultrasound
  (d) Computed tomography (CT) scans

Based on this information, the PT can make a differential diagnosis and rehabilitation plan.

During the rehabilitation process, the PT will conduct performance assessments to determine the progression of the rehabilitation. The location and type of injury determine the performance assessment the PT will conduct. Many resources exist that explain these types of assessments [22, 23].

Conclusion

In conclusion, the human performance optimization program's musculoskeletal domain crucially depends on the integrated and collaborative efforts of exercise physiologists,strength coaches, athletic trainers (ATCs), and physical therapists. Each professional brings unique expertise, perspectives, and treatment methodologies to address musculoskeletal health and performance optimization.

Exercise physiologists play a vital role in developing and managing conditioning programs, providing their expertise in exercise science to enhance athletes' performance and prevent injury. Their understanding of human metabolism, physiology, and biomechanics enables them to design individualized, evidenced-based training programs that promote optimal physical functioning.

On the other hand, athletic trainers (ATCs) have specialized skills in preventing, recognizing, managing, and rehabilitating injuries that result from physical activity. Their immediate and ongoing care ensures that athletes receive appropriate injury prevention measures and rehabilitation, leading to a quicker, safer return to play.

With their specialized knowledge of the human musculoskeletal system, physical therapists provide vital intervention in cases of injury or impairment. They offer treatment to alleviate pain and restore function and educate athletes on how to manage their conditions and prevent future injuries.

Together, these professionals form a powerful triad of care within the MSK Domain, demonstrating the critical importance of interprofessional collaboration in human performance programs. Their distinct yet interrelated roles underscore the need for ongoing communication, education, and mutual respect among these professions. As HPO programs expand in knowledge and technology, the professional composition within this domain will inevitably evolve. Yet the common goal will always remain: to optimize human performance and promote the health and well-being of all HWS professionals.

References

1. Exercise Physiologist [Online]. [date unknown]. https://www.acsm.org/certification/get-certified/exercise-physiologist?gclid=Cj0KCQjwxMmhBhDJARIsANFGOSt89rxu8T4BDCDCABjLDroIbp3szv9ZAlZZLFMhlORW2FPwl0TdUb0aAmo3EALw_wcB. [9 Apr. 2023].
2. Exercise Physiologists: Occupational Outlook Handbook. U.S. Bureau of Labor Statistics [Online]. [date unknown]. https://www.bls.gov/ooh/healthcare/exercise-physiologists.htm. [9 Apr 2023].
3. Department of the Air Force Instruction 44-171: Medical Air Force Medical Home Clinical Operations [Online]. http://www.e-publishing.af.mil/http://www.e-Publishing.af.mil/.
4. Become a Certified Strength & Conditioning Specialist | NSCA [Online]. [date unknown]. https://www.nsca.com/certification/cscs/?gclid=Cj0KCQjwxMmhBhDJARIsANFGOSsV0ZS6OgmFd7fJKF5ctw1w-19zME9cc01HkxdpIUdJARPFMct-lzEaAp5SEALw_wcB. [9 Apr 2023].
5. Becoming a Physical Therapist | APTA [Online]. [date unknown]. https://www.apta.org/your-career/careers-in-physical-therapy/becoming-a-pt. [9 Apr 2023].
6. Long P. Athletic trainer's in the military setting. 2023.
7. Membership | NATA [Online]. [date unknown]. https://www.nata.org/membership?gclid=Cj0KCQjwxMmhBhDJARIsANFGOStpdGeV_UKDhMQFqp0T59xK9X0iuss38igq9_Xm-ivzoOWHLHBGqrkaAsBDEALw_wcB. [9 Apr 2023].
8. Callender A. The Sports Medicine Licensure Clarity Act | NATA [Online]. 2021. https://www.nata.org/sports-medicine-licensure-clarity-act. [9 Apr 2023].
9. Defense Health Agency. DHA procedure manual 6010.13. 2008.
10. Becoming a Physical Therapist Assistant | APTA [Online]. [date unknown]. https://www.apta.org/your-career/careers-in-physical-therapy/becoming-a-pta. [9 Apr 2023].
11. Haff G, Triplett NT. Essentials of strength training and conditioning. 2014.
12. Brown L, editor. Strength training [Online]. 2nd ed. Human Kinetics. https://www.nsca.com/store/product-detail/INV/9781492522089/9781492522089. [9 Apr 2023].
13. Gunn S, Mann B, Kazadi K. Developing explosive athletes: use of velocity based training in athletes [Online]. 4th ed. https://www.amazon.de/-/en/Steve-Gunn/dp/

B095GLNSVQ/ref=pd_lpo_1?pd_rd_w=MvgcD&content-id=amzn1.sym.992f08af-eb2a-4a7b-90a4-cd77032caf05&pf_rd_p=992f08af-eb2a-4a7b-90a4-cd77032caf05&pf_rd_r=AK50CJJ1GDEECDJ8DRJP&pd_rd_wg=wJUOs&pd_rd_r=59a7ed08-bf58-496a-8a. [9 Apr 2023].

14. Signore N. Velocity-based training: how to apply science, technology, and data to maximize performance [Online]. 1st ed. Human Kinetics. https://www.amazon.de/-/en/Nunzio-Signore/dp/1492599956. [9 Apr 2023].
15. McGuigan M, editor. Developing power. 1st ed. Human Kinetics: Champaign; 2017.
16. Department of the Air Force. DAFMAN 36-2905. .
17. Myers C, Allen H. Triathlon training with power [Online]. 1st ed. Cognella Publishing. https://titles.cognella.com/triathlon-training-with-power-9781793532725. [11 Sept 2021].
18. Tanner RK, Gore CJ. Physiological tests for elite athletes. 2nd ed. 2013.
19. Reuter B, editor. Developing endurance. 1st ed. Human Kinetics: Champaign; 2012.
20. Jeffreys I, editor. Developing speed. 1st ed. Human Kinetics: Champaign; 2013.
21. Dawes J, editor. Developing agility and quickness. 2nd ed. Human Kinetics: Champaign; 2019.
22. Brukner P, Khan K, Clarsen B, Cook J (Professor), Cools A, Crossley K, Hutchinson MR, McCrory P, Bahr R, Brukner P. Brukner & Khan's clinical sports medicine: Volume 1, Injuries. McGraw Hill; 2017.
23. Brukner P, Khan K. Brukner Khan's Clinical sports medicine. Volume 2, The medicine of exercise. McGraw-Hill; 2019.

Chapter 5
The Nutrition Domain

Christopher Myers and Elizabeth Foley

Introduction

Human weapon system (HWS) optimization cannot occur without nutrition. Good nutrition enhances human performance and can help improve psychological and cognitive performance (Fig. 5.1) [1–3]. Often, nutrition programs focus on fueling strategies that are designed to optimize performance. However, this is only sometimes the case for the HWS. Most HWS professions do not need these types of nutritional strategies. Instead, these professionals should have nutrition strategies focusing on eating for health rather than peak performance. By striving for healthier, balanced foodstuff consumption, nutrition can be a powerful tool for preventative healthcare. The Nutrition Domain is the bridge to optimizing the MSK, Cognitive, and Psychological Domains.

Furthermore, individual variability in the optimal diet recognizes the unique interplay between a person's genetics, gut microbiome, lifestyle, and environmental factors and how these influence dietary needs and responses. For example, some individuals may have a genetic disposition to metabolize fats efficiently and thus may perform better on a higher-fat diet. In contrast, others may thrive on a diet rich in complex carbohydrates.

Moreover, the gut microbiome – the diverse collection of microbes in our digestive system – influences our health and well-being, including our nutritional needs and how we respond to different foods. Individual differences in gut microbiota may explain why some people react differently to specific diets.

This individual variability extends beyond macronutrients. Certain individuals may require higher levels of specific micronutrients due to genetic variants affecting

C. Myers (✉) · E. Foley
3 Sports Science, Founder and Head Coach, Akron, OH, USA
e-mail: chris@3sportsscience.com

C. Myers, *The Human Weapon System*,
https://doi.org/10.1007/978-3-031-45061-7_5

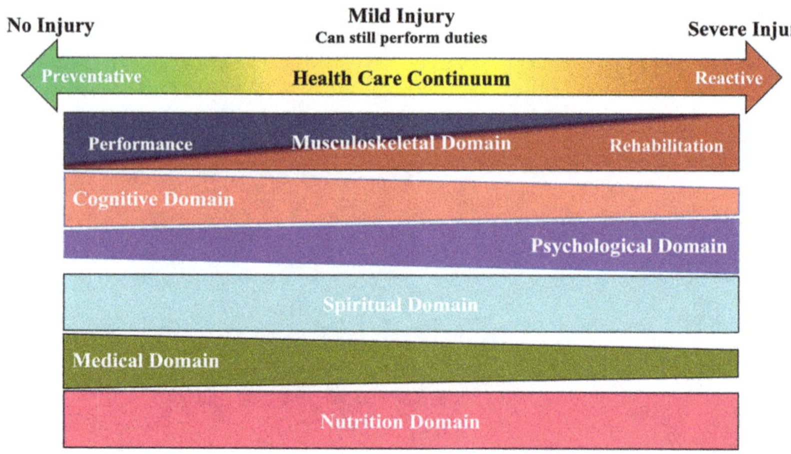

Fig. 5.1 The Nutrition Domain supports the other HPO domains in the Healthcare Continuum (same as Fig. 3.7)

their metabolism, absorption, or utilization of these nutrients. Understanding this individual variability in dietary needs and responses drives a shift toward personalized nutrition in the performance nutrition landscape. Registered dieticians (RDs) provide this expertise within the human performance optimization (HPO) program. Customized food strategies consider an individual's unique genetic, physiological, and lifestyle factors to tailor dietary recommendations specifically for them. The goal is to optimize their dietary intake for enhanced performance, faster recovery, improved health outcomes, and overall well-being.

The Nutrition Domain team provides highly individualized dietary recommendations by harnessing genomics, metabolomics, and gut microbiome research advancements. This approach promises to enhance athletic performance, promote long-term health, and prevent disease, marking an exciting new direction for performance nutrition.

Nutrition as Prevention

While nutrition for optimizing human performance is often thought of strictly as physical activity, the importance of using nutrition to prevent noncommunicable diseases cannot be overstated. Beyond nutrition's ability to optimize performance during intense physical activity, nutrition is pivotal in preventing chronic disease. Unhealthy lifestyle behaviors account for the vast majority of noncommunicable diseases and are the culprit behind about 63% of all deaths [4].

Dietary Patterns and Inflammation

Inflammation is often a hallmark sign of chronic disease, aging, and a general decrease in long-term health. While acute inflammation can be healthy and aid in healing (i.e., inflammation after physical activity or an acute trauma), chronic inflammation can lead to long-term adverse health effects. Indeed, a review by Furman et al. investigating chronic inflammation in the etiology of disease across the life span lists an increased risk of metabolic syndrome, type 2 diabetes, nonalcoholic fatty liver disease (NAFLD), hypertension, cardiovascular disease (CVD), chronic kidney disease, various types of cancer, depression, neurodegenerative and autoimmune diseases, osteoporosis, and sarcopenia [5]. It is well established that inflammation contributes to the onset and progression of diseases like metabolic syndrome, type 2 diabetes, and CVD. Inflammation is also associated with multiple mental health disorders, most notably major depressive disorder (MDD) and post-traumatic stress disorder (PTSD); individuals with these disorders have significantly elevated blood levels of inflammatory markers, such as IL-1β, IL-6, TNF-α, and CRP [6]. While it is unknown whether this increase in proinflammatory biomarkers comes before or after the onset of either disorder, the fact remains that each of the biomarkers mentioned above is also associated with chronic disease states. Indeed, PTSD is associated with significantly increased physical comorbidities such as metabolic syndrome, CVD, and autoimmune disease [6]. While this short background on inflammation, chronic disease, and even mental health disorders may seem entirely unrelated to nutrition, it is well established that diet and other lifestyle factors significantly impact inflammation, chronic illness, and mental health. Proper regulation of inflammatory processes is of paramount importance in the HWS due to its inherent role in the onset or progression of different chronic diseases and mental health disorders, and therefore, nutrient deficiencies, excesses, and general patterns are as well.

Dietary Patterns and Chronic Disease

Dietary patterns are some of the most important predictors of future health. The Mediterranean diet is hailed as one of the healthiest dietary patterns due to the antioxidants and anti-inflammatory nutrients present in this mixture of food; this particular dietary pattern focuses on increasing the consumption of omega-3 fatty acids, legumes, nuts, fish, fruits, and vegetables. A meta-analysis by Garcia-Hermoso et al. found a positive association between adherence to a Mediterranean diet and physical activity, cardiorespiratory fitness, and muscular fitness [7]. Other studies have found that the Mediterranean diet decreases the risk of developing diabetes [8], metabolic disorders [9], mental health diseases like depression [10], and cognitive decline [11]. The Mediterranean diet is not overly restrictive and allows for various foods and tastes, thus making it an excellent option for individuals who are already fit and trying to prevent/treat chronic disease.

Conversely, another newly popular dietary pattern focuses on time-restricted eating rather than specific dietary nutrients. Time-restricted eating limits an individual's feeding period throughout the day; a person's fast can be anywhere from 12 to 21 hours per day, with a feeding window anywhere from 3 to 12 h per day. Understandably, there is concern that this eating pattern could lead to decreased performance. Still, a 2016 study in resistance-trained males found that restricting their feeding schedule to a period of 8 h resulted in a reduction of fat mass without compromising strength [12]. Another study on male cyclists found that time-restricted feeding decreases inflammation, increases performance output, and improves body composition [13]. Tinsley et al. also found that while time-restricted feeding tended to reduce caloric intake inadvertently, resistance-trained subjects still saw increased muscular and endurance performance [14]. A systematic review by Adafer et al. found that time-restricted eating was metabolically beneficial with or without fat loss and that fat loss took place regardless of caloric restriction [15]. These authors suggested that this was due to an intrinsic effect of realigning the feeding schedule around circadian rhythm. This important factor should be considered when attempting to optimize the HWS.

Another popular dietary pattern that has gained traction recently includes the ketogenic diet. The ketogenic diet is characterized by high amounts of fat, extremely low carbohydrates, and adequate protein levels. While the ketogenic diet was initially recommended to decrease seizures in patients with epilepsy [16], it has gained popularity for weight loss [17]. While this diet has undoubtedly been shown to benefit individuals with chronic diseases like type 2 diabetes [18], migraines [19], and cardiometabolic disorders [20–22], its utility for optimizing performance is up for debate. A study by Volek et al. found that ultra-endurance-trained runners who consumed a ketogenic diet for an average of 20 months displayed a significantly increased aptitude to oxidate fat during endurance exercise while maintaining normal skeletal muscle glycogen [23]. While another study by Burke et al. [24] also reported an increase in participants' ability to oxidize fat, the participants consuming a ketogenic diet underperformed compared to participants consuming carbohydrates; the ketogenic group also reported an increase in perceived effort and perceived workout quality. It is important to note that this study took place over 4 weeks, suggesting that more time may be necessary for adaptations to improve training and performance [24]. Additionally, a review by Raimondo et al. noted that while the ketogenic diet consistently decreases fat mass and improves fatty acid oxidation capabilities, there was little to no positive benefit regarding performance [25]. While ketones are shown to be more efficient at generating metabolic energy than glucose, the feasibility of the diet itself, the acute drop in performance, and the perceived exertion of participants all need to be taken into account when considering the ketogenic diet as a method of optimizing the HWS [26].

Alcohol

While carbohydrates, proteins, and lipids are the well-known essential macronutrients, there is a fourth: alcohol. Alcohol is a commonly used substance whose health risks and benefits remain widely debated. Public health officials view alcohol as a significant public health issue, going so far as to say that the safest amount of alcohol to consume is 0 g of alcohol [27]. Alcohol is associated with a significant risk of developing cancer [28], risk of developing age-related macular degeneration [29], and liver cirrhosis [30]. However, the moderate consumption of red wine in the Mediterranean diet is associated with cardiometabolic protective effects [31]. Average alcohol consumption often has an inverse relationship with cardiovascular disease risk [32]. Still, it has no other benefits and increases cancer risk and other diseases [31, 33]. Alcohol also increases intestinal inflammation [34] as well as neural inflammation [35], which can lead to detrimental systemic effects and is a known sleep disruptor [36].

Though alcohol consumption comes with known risk factors and if overconsumption can result in acute discomfort as a "hangover," it is still a widely used substance. Indeed, more physically active individuals report higher alcohol use [37, 38], which begs the question: How does alcohol affect performance and recovery? A study on recreationally active individuals that measured exercise performance after previous-day alcohol use found that aerobic performance was significantly decreased by alcohol consumption, but resistance performance measures remained unchanged [39]. Conversely, another study investigating the effect of not drinking alcohol for 4 weeks on cardiorespiratory fitness did not see any changes when comparing alcohol abstinence to alcohol use. These conflicting results may partly be due to when performance parameters were measured, i.e., measuring the performance of someone who drinks moderately but did not drink the night prior versus measuring the performance of someone who partied the night before. Though its effect on performance seems to vary, alcohol has been consistently found to decrease muscle synthesis [40] due to altered mTOR signaling [41], a suppressive effect on protein synthesis even when used alongside carbohydrates and protein [42]. Regarding optimizing the HWS, alcohol is not a macronutrient that is necessary or recommended for consumption.

The Nutrition Domain

The Nutrition Domain is part of the MSK Domain (Fig. 5.3). Proper nutrition is essential for the HWS to reach peak performance [43–48]. Accurately fueling the HWS improves workout quality and recovery. Better recovery times mean the HWS can do more sets and repetitions during workouts, enhancing physical performance.

The Nutrition team works in tandem with the MSK teams. The nutritional expertise of the nutrition team supports the HWS performance and rehabilitation processes. A cohesive MSK and Nutrition team provides a synergy and level of optimization not seen when these domains work individually (Fig. 5.3).

Figures 5.1 and 5.2 may sound contradictory; however, the Nutrition Domain supports all the domains in the HPO paradigm within the HC continuum. The order shown in Figs. 3.7 and 5.1 matters. Proper nutrition helps other aspects of HPO (Fig. 5.1).

Proper nutrition is shown to sustain and improve psychological and cognitive performance. However, recent research has shown that suboptimal nutrition can create pathology in several behavioral health disorders, such as anxiety and depression [49, 50]. This pathology occurs through the disruption of neurotransmitters, the hypothalamus-pituitary gland-adrenal axis, an increase in oxidative stress and inflammation, reductions in neuroplasticity, and reductions in mitochondrial function [49]. Controlled randomized trials in humans have shown that proper nutrition can improve symptoms of depression and anxiety [50–55]. Since adequate nutrition supports all elements of HPO (Fig. 5.1), the Nutrition Domain has its domain. This mutually inclusive aspect between all the fields optimizes human performance (Fig. 5.4). Remember, this synergy is not properly optimizing a human performance program. An HPO program must still have the HWS buy-in component to be truly optimized.

The Nutrition Domain Occupation Composition

A few different occupations can round out the Nutrition Domain team. However, depending on the size of the HWS unit or organization and funding, the Nutrition Domain may only be staffed by a dietitian. A dietitian is an allied health professional who is a credentialed food expert who can prescribe dietary treatments and perform dietary counseling for a wide range of medical and allergy issues to improve physical performance [56–58].

A registered dietitian (RD) must have at least a master's degree from a university accredited by Accreditation Council for Education in Nutrition and Dietetics

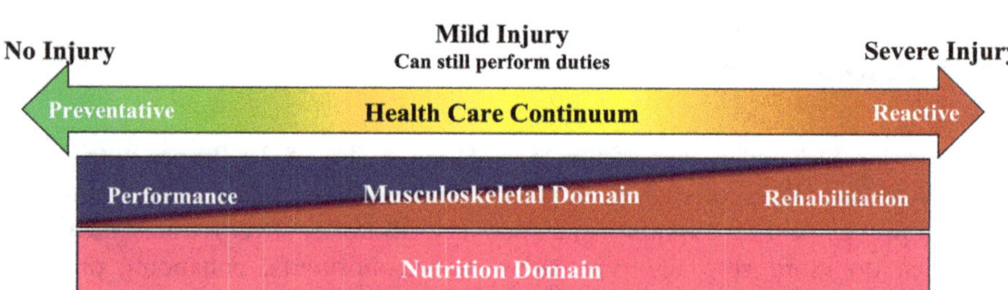

Fig. 5.2 The nutrition domain supports the MSK domain concerning the Healthcare Continuum

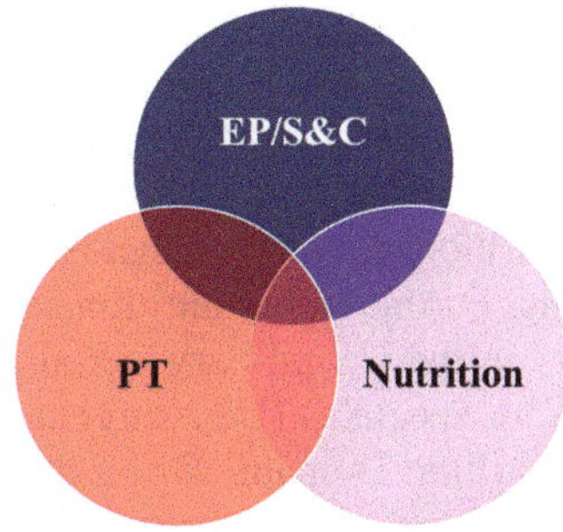

Fig. 5.3 This figure depicts how the MSK domain teams work in synergy with the nutrition domain team

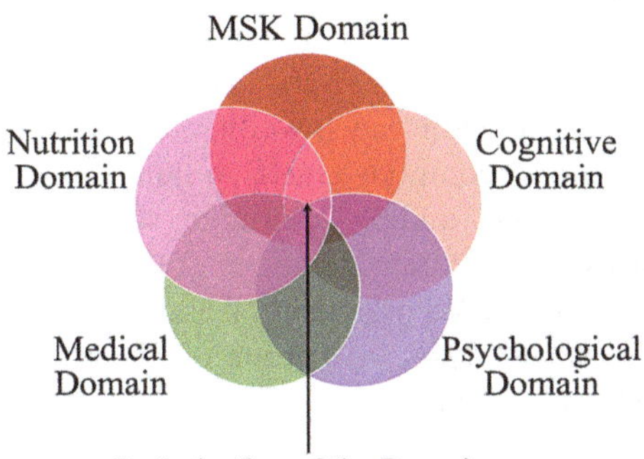

Fig. 5.4 This figure depicts the synergy between all the domains. The center of this diagram is the point where true synergy occurs. However, this synergy must be appropriately optimized for a human performance program. For example, an HPO program must still have the HWS buy-in component to be truly optimized

(ACEND), complete a supervised practicum, pass a state licensure exam, and participate in continuing education programs [59]. RDs can be considered privileged providers, but this aspect is determined by the supervising hospital or the local health organization [60].

A nutritionist is an entirely separate profession from an RD. No clear standard exists for nutritionists. Anyone who has taken a sports nutrition class or basic sports nutrition certification can call themselves a nutritionist. Within the field, some informal criteria are used to discern the quality of nutritionists. Many organizations look for a nutritionist who has completed at least a bachelor's degree in exercise science or a similar field with an element of nutrition courses with a certification in sports nutrition from a recognized organization, such as the International Society of Sports Nutrition (ISSN). The academic program does not need to come from an accredited university. However, having an education from an accredited university assures a baseline of knowledge recognized to be required to be successful in the field. Finally, many organizations want nutritionists to have 2–3 years of experience in the area.

A significant difference exists in what nutritionists can do compared to RDs. This difference is due to the education and credentialing differences between the two professions. Nutritionists are not considered allied health professionals or regarded as privileged providers. Nutritionists are limited to teaching general nutrition and the relationship between food and health. In some states, based on local laws, nutritionists may be able to focus on food behaviors. These limitations mean nutritionists cannot diagnose or treat medical conditions or provide dietary counseling. Furthermore, nutritionists cannot prescribe meal plans [56, 61].

In the US military, certain military occupational specialties, such as Nutrition Care Specialist (US Army) and Diet Technician (US Air Force) exist to augment the Nutrition Domain. These specialties are essentially RD assistants. These occupational specialties work under the supervision of an RD. Due to their scope of work, nutrition care specialists and diet technicians have more responsibility than those in the civilian sector. For example, these occupational specialties can execute meal plans and nutritional therapies for military service members prescribed by the RD. Additionally, these occupations teach wellness and nutrition [44, 62].

The Nutrition Domain can comprise an RD, a nutritionist, and an RD assistant. The size of the HWS unit and funding determines the size of the nutrition team. Fundamental differences exist in the capabilities of the professions associated with this domain. The needs of the HWS organization and the goals of the HPO program, as determined by the needs analysis, will guide the determination of the composition of the Nutrition Domain team.

Performance Metrics of the Nutrition Domain

Optimizing human performance demands a comprehensive understanding of multiple factors, among which nutrition stands paramount. As such, dieticians play a crucial role in any human performance optimization program. They employ various metrics to evaluate an individual's dietary habits, overall health, and potential nutritional interventions. Here, we explore some key metrics that dieticians typically use in the context of performance optimization:

1. Dietary intake analysis: dieticians often start by thoroughly evaluating an individual's dietary intake, including the types and amounts of food and beverages consumed. This assessment may involve using dietary recall surveys, food frequency questionnaires, or food diaries to understand the individual's typical consumption patterns. These questionnaires and surveys typically ask clients to recall their meal and dietary habits for 1, 3, or 7 days. The analysis can provide insights into macronutrient and micronutrient intake and help identify areas where nutritional enhancements can be made [48].
2. Body composition analysis: understanding an individual's body composition, including levels of fat mass and lean body mass, is a vital part of performance optimization. Techniques such as bioelectrical impedance analysis (BIA), skinfold measurements, air displacement plethysmography, or dual-energy X-ray absorptiometry (DXA) can be employed. By monitoring body composition over time, dieticians can assess the effectiveness of dietary interventions in achieving desired body composition outcomes [63–70].

3. Anthropometric measurements: these measurements include body weight, height, and girths of various body segments. Together with body composition analysis, these metrics can help the dietician monitor changes over time, establish nutritional needs, and set appropriate goals [63–66].
4. Energy expenditure estimation: dieticians assess total energy expenditure (TEE), which includes basal metabolic rate (BMR), physical activity level (PAL), and the thermic effect of food (TEF). Tools such as metabolic carts, accelerometers, or predictive equations (e.g., Harris-Benedict, Mifflin-St. Jeor) can be used. This estimation helps tailor dietary plans that balance energy intake with expenditure, which is crucial for weight management and performance.
5. Biochemical data: blood, urine, or other biological samples can provide critical information about an individual's nutritional status, hydration level, and potential deficiencies. For example, blood tests might reveal electrolyte imbalances or inadequate levels of vitamins or minerals [63–66].
6. Performance metrics: dieticians also look at performance data, such as strength, endurance, recovery time, and others. They collaborate with other professionals (like exercise physiologists and athletic trainers) to understand how diet affects these metrics and make evidence-based recommendations for nutritional interventions [48, 71].

By systematically using these metrics, the Nutrition Domain team can provide targeted, personalized nutritional strategies to optimize human performance. The goal is to enhance an individual's health, well-being, and performance through the most powerful yet often overlooked tool: nutrition.

Conclusion

Nutrition is more than just fueling to perform; it is a way to prevent disease and improve behavioral health. This domain provides a foundation for all other domains to improve health and fitness along the Healthcare Continuum. The HWS organization composition, HPO program goals, and funding determine the design of a Nutritional Domain team. The type of nutrition professionals on the team determines the capability of the Nutrition Domain for the HWS organization. With proper nutrition, HWS can be optimized.

References

1. Martin N. Performance nutrition: the New Zealand Defence force perspective. J Sci Med Sport. 2017;20 https://doi.org/10.1016/j.jsams.2017.09.515.
2. Gonzalez DE, McAllister MJ, Waldman HS, Ferrando AA, Joyce J, Barringer ND, Dawes JJ, Kieffer AJ, Harvey T, Kerksick CM, Stout JR, Ziegenfuss TN, Zapp A, Tartar JL, Heileson JL, VanDusseldorp TA, Kalman DS, Campbell BI, Antonio J, Kreider RB. International society of sports nutrition position stand: tactical athlete nutrition. J Int Soc Sports Nutr. 2022;19 https://doi.org/10.1080/15502783.2022.2086017.

3. Lieberman HR. Nutrition, brain function and cognitive performance. Appetite. 2003;40:245–54.
4. Kushner RF, Sorensen KW. Lifestyle medicine. Curr Opin Endocrinol Diabetes Obes. 2013;20:389–95. https://doi.org/10.1097/01.med.0000433056.76699.5d.
5. Furman D, Campisi J, Verdin E, Carrera-Bastos P, Targ S, Franceschi C, Ferrucci L, Gilroy DW, Fasano A, Miller GW, Miller AH, Mantovani A, Weyand CM, Barzilai N, Goronzy JJ, Rando TA, Effros RB, Lucia A, Kleinstreuer N, Slavich GM. Chronic inflammation in the etiology of disease across the life span. Nat Med. 2019;25:1822–32. https://doi.org/10.1038/s41591-019-0675-0.
6. Hori H, Kim Y. Inflammation and post-traumatic stress disorder. Psychiatry Clin Neurosci. 2019;73:143–53. https://doi.org/10.1111/pcn.12820.
7. García-Hermoso A, Ezzatvar Y, López-Gil JF, Ramírez-Vélez R, Olloquequi J, Izquierdo M. Is adherence to the Mediterranean diet associated with healthy habits and physical fitness? A systematic review and meta-analysis including 565 421 youths. Br J Nutr. 2022;128:1433–44. https://doi.org/10.1017/S0007114520004894.
8. Becerra-Tomás N, Blanco Mejía S, Viguiliouk E, Khan T, Kendall CWC, Kahleova H, Rahelić D, Sievenpiper JL, Salas-Salvadó J. Mediterranean diet, cardiovascular disease and mortality in diabetes: a systematic review and meta-analysis of prospective cohort studies and randomized clinical trials. Crit Rev Food Sci Nutr. 2020;60:1207–27. https://doi.org/10.1080/10408398.2019.1565281.
9. Godos J, Zappalà G, Bernardini S, Giambini I, Bes-Rastrollo M, Martinez-Gonzalez M. Adherence to the Mediterranean diet is inversely associated with metabolic syndrome occurrence: a meta-analysis of observational studies. Int J Food Sci Nutr. 2017;68:138–48. https://doi.org/10.1080/09637486.2016.1221900.
10. Pagliai G, Sofi F, Vannetti F, Caiani S, Pasquini G, Molino Lova R, Cecchi F, Sorbi S, Macchi C. Mediterranean diet, food consumption and risk of late-life depression: the Mugello study. J Nutr Health Aging. 2018;22:569–74. https://doi.org/10.1007/s12603-018-1019-3.
11. Aridi YS, Walker JL, Wright ORL. The association between the Mediterranean dietary pattern and cognitive health: a systematic review. Nutrients. 2017;9 https://doi.org/10.3390/nu9070674.
12. Moro T, Tinsley G, Bianco A, Marcolin G, Pacelli QF, Battaglia G, Palma A, Gentil P, Neri M, Paoli A. Effects of eight weeks of time-restricted feeding (16/8) on basal metabolism, maximal strength, body composition, inflammation, and cardiovascular risk factors in resistance-trained males. J Transl Med. 2016;14:290. https://doi.org/10.1186/s12967-016-1044-0.
13. Moro T, Tinsley G, Longo G, Grigoletto D, Bianco A, Ferraris C, Guglielmetti M, Veneto A, Tagliabue A, Marcolin G, Paoli A. Time-restricted eating effects on performance, immune function, and body composition in elite cyclists: a randomized controlled trial. J Int Soc Sports Nutr. 2020;17:65. https://doi.org/10.1186/s12970-020-00396-z.
14. Tinsley GM, Forsse JS, Butler NK, Paoli A, Bane AA, La Bounty PM, Morgan GB, Grandjean PW. Time-restricted feeding in young men performing resistance training: a randomized controlled trial. Eur J Sport Sci. 2017;17:200–7. https://doi.org/10.1080/17461391.2016.1223173.
15. Adafer R, Messaadi W, Meddahi M, Patey A, Haderbache A, Bayen S, Messaadi N. Food timing, circadian rhythm and chrononutrition: a systematic review of time-restricted eating's effects on human health. Nutrients. 2020;12 https://doi.org/10.3390/nu12123770.
16. Barzegar M, Afghan M, Tarmahi V, Behtari M, Rahimi Khamaneh S, Raeisi S. Ketogenic diet: overview, types, and possible anti-seizure mechanisms. Nutr Neurosci. 2021;24:307–16. https://doi.org/10.1080/1028415X.2019.1627769.
17. Basolo A, Magno S, Santini F, Ceccarini G. Ketogenic diet and weight loss: is there an effect on energy expenditure? Nutrients. 2022;14:1814. https://doi.org/10.3390/nu14091814.
18. Zhou C, Wang M, Liang J, He G, Chen N. Ketogenic diet benefits to weight loss, glycemic control, and lipid profiles in overweight patients with type 2 diabetes mellitus: a meta-analysis of randomized controlled trails. Int J Environ Res Public Health. 2022;19 https://doi.org/10.3390/ijerph191610429.

19. Bongiovanni D, Benedetto C, Corvisieri S, Del Favero C, Orlandi F, Allais G, Sinigaglia S, Fadda M. Effectiveness of ketogenic diet in treatment of patients with refractory chronic migraine. Neurol Sci. 2021;42:3865–70. https://doi.org/10.1007/s10072-021-05078-5.
20. Sharman MJ, Kraemer WJ, Love DM, Avery NG, Gómez AL, Scheett TP, Volek JS. A ketogenic diet favorably affects serum biomarkers for cardiovascular disease in normal-weight men. J Nutr. 2002;132:1879–85. https://doi.org/10.1093/jn/132.7.1879.
21. Cicero AFG, Benelli M, Brancaleoni M, Dainelli G, Merlini D, Negri R. Middle and long-term impact of a very low-carbohydrate ketogenic diet on cardiometabolic factors: a multi-center, cross-sectional, clinical study. High Blood Press Cardiovasc Prev. 2015;22:389–94. https://doi.org/10.1007/s40292-015-0096-1.
22. Gershuni VM, Yan SL, Medici V. Nutritional ketosis for weight management and reversal of metabolic syndrome. Curr Nutr Rep. 2018;7:97–106. https://doi.org/10.1007/s13668-018-0235-0.
23. Volek JS, Freidenreich DJ, Saenz C, Kunces LJ, Creighton BC, Bartley JM, Davitt PM, Munoz CX, Anderson JM, Maresh CM, Lee EC, Schuenke MD, Aerni G, Kraemer WJ, Phinney SD. Metabolic characteristics of keto-adapted ultra-endurance runners. Metabolism. 2016;65:100–10. https://doi.org/10.1016/j.metabol.2015.10.028.
24. Burke LM, Ross ML, Garvican-Lewis LA, Welvaert M, Heikura IA, Forbes SG, Mirtschin JG, Cato LE, Strobel N, Sharma AP, Hawley JA. Low carbohydrate, high fat diet impairs exercise economy and negates the performance benefit from intensified training in elite race walkers. J Physiol. 2017;595:2785–807. https://doi.org/10.1113/JP273230.
25. Di Raimondo D, Buscemi S, Musiari G, Rizzo G, Pirera E, Corleo D, Pinto A, Tuttolomondo A. Ketogenic diet, physical activity, and hypertension – a narrative review. Nutrients. 2021;13 https://doi.org/10.3390/nu13082567.
26. Veech RL. The therapeutic implications of ketone bodies: the effects of ketone bodies in pathological conditions: ketosis, ketogenic diet, redox states, insulin resistance, and mitochondrial metabolism. Prostaglandins Leukot Essent Fatty Acids. 2004;70:309–19. https://doi.org/10.1016/j.plefa.2003.09.007.
27. Burton R, Sheron N. No level of alcohol consumption improves health. Lancet. 2018;392:987–8. https://doi.org/10.1016/S0140-6736(18)31571-X.
28. Rumgay H, Shield K, Charvat H, Ferrari P, Sornpaisarn B, Obot I, Islami F, Lemmens VEPP, Rehm J, Soerjomataram I. Global burden of cancer in 2020 attributable to alcohol consumption: a population-based study. Lancet Oncol. 2021;22:1071–80. https://doi.org/10.1016/S1470-2045(21)00279-5.
29. Zhang J, Mitsuhashi T, Matsuo T, Yorifuji T, Hamada J, Liu Y. Alcohol consumption and age-related macular degeneration: a systematic review and dose–response meta-analysis. Curr Eye Res. 2021;46:1900–7. https://doi.org/10.1080/02713683.2021.1942070.
30. Ma H, Li X, Zhou T, Sun D, Shai I, Heianza Y, Rimm EB, Manson JE, Qi L. Alcohol consumption levels as compared with drinking habits in predicting all-cause mortality and cause-specific mortality in current drinkers. Mayo Clin Proc. 2021;96:1758–69. https://doi.org/10.1016/j.mayocp.2021.02.011.
31. Martínez-González MA, Sánchez-Villegas A. The emerging role of Mediterranean diets in cardiovascular epidemiology: monounsaturated fats, olive oil, red wine or the whole pattern? Eur J Epidemiol. 2004;19:9–13. https://doi.org/10.1023/b:ejep.0000013351.60227.7b.
32. Klatsky AL. Alcohol and cardiovascular mortality: common sense and scientific truth. J Am Coll Cardiol. 2010;55:1336–8. https://doi.org/10.1016/j.jacc.2009.10.057.
33. Barbería-Latasa M, Gea A, Martínez-González MA. Alcohol, drinking pattern, and chronic disease. Nutrients. 2022;14 https://doi.org/10.3390/nu14091954.
34. Bishehsari F, Magno E, Swanson G, Desai V, Voigt RM, Forsyth CB, Keshavarzian A. Alcohol and gut-derived Inflammation. [Online]. Alcohol Res. 2017;38:163–71. http://www.ncbi.nlm.nih.gov/pubmed/28988571. 5 Sept 2023
35. Palmer E, Tyacke R, Sastre M, Lingford-Hughes A, Nutt D, Ward RJ. Alcohol hangover: underlying biochemical, inflammatory and neurochemical mechanisms. Alcohol Alcohol. 2019;54:196–203. https://doi.org/10.1093/alcalc/agz016.

36. Colrain IM, Nicholas CL, Baker FC. Alcohol and the sleeping brain. Handb Clin Neurol. 2014;125:415–31. https://doi.org/10.1016/B978-0-444-62619-6.00024-0.
37. Henderson CE, Najjar LZ, Young CM, Leasure JL, Neighbors C, Gasser ML, Lindgren KP. Longitudinal relations between physical activity and alcohol consumption among young adults. Psychol Addict Behav. 2023;37:285–93. https://doi.org/10.1037/adb0000807.
38. Dodge T, Clarke P, Dwan R. The relationship between physical activity and alcohol use among adults in the United States. Am J Health Promot. 2017;31:97–108. https://doi.org/10.1177/0890117116664710.
39. Shaw AG, Chae S, Levitt DE, Nicholson JL, Vingren JL, Hill DW. Effect of previous-day alcohol ingestion on muscle function and performance of severe-intensity exercise. Int J Sports Physiol Perform. 2022;17:44–9. https://doi.org/10.1123/ijspp.2020-0790.
40. Korzick DH, Sharda DR, Pruznak AM, Lang CH. Aging accentuates alcohol-induced decrease in protein synthesis in gastrocnemius. Am J Physiol Regul Integr Comp Physiol. 2013;304:R887–98. https://doi.org/10.1152/ajpregu.00083.2013.
41. Levitt DE, Luk H-Y, Vingren JL. Alcohol, resistance exercise, and mTOR pathway signaling: an evidence-based narrative review. Biomol Ther. 2022;13 https://doi.org/10.3390/biom13010002.
42. Parr EB, Camera DM, Areta JL, Burke LM, Phillips SM, Hawley JA, Coffey VG. Alcohol ingestion impairs maximal post-exercise rates of myofibrillar protein synthesis following a single bout of concurrent training. PLoS One. 2014;9:e88384. https://doi.org/10.1371/journal.pone.0088384.
43. USADA. Nutrition guide: handbook of nutritional contents [Online]. http://www.truesport.org
44. Nutritional Fitness [Online]. [date unknown]. https://www.airforcemedicine.af.mil/Resources/Health-Promotion/Nutritional-Fitness/. 15 Apr 2023.
45. Ormsbee MJ, Bach CW, Baur DA. Pre-exercise nutrition: the role of macronutrients, modified starches and supplements on metabolism and endurance performance. Nutrients. 2014;6:1782–808. https://doi.org/10.3390/nu6051782.
46. Marriott BM (Institute of M (US) C on MNR, Carlson SJI of M (US) C on MNR. Nutritional needs in cold and high-altitude environments. 1996.
47. Smith SM, Davis-street JE, Fesperman JV, Smith MD, Rice BL, Zwart SR. Human nutrition and metabolism nutritional status changes in humans during a 14-day saturation dive: the NASA Extreme Environment Mission Operations V Project 1, 2. J Nutr. 2004:1765–71.
48. Jeukendrup AE, Gleeson M. Sport nutrition [Online]. https://www.amazon.de/Sport-Nutrition-Asker-Jeukendrup/dp/1492529036/ref=asc_df_1492529036/?tag=googshopde-21&linkCode=df0&hvadid=310619028235&hvpos=&hvnetw=g&hvrand=13307388785746463045&hvpone=&hvptwo=&hvqmt=&hvdev=c&hvdvcmdl=&hvlocint=&hvlocphy=9042720&hvtargid=pla-524500659240&psc=1&th=1&psc=1&tag=&ref=&adgrpid=65539873561&hvpone=&hvptwo=&hvadid=310619028235&hvpos=&hvnetw=g&hvrand=13307388785746463045&hvqmt=&hvdev=c&hvdvcmdl=&hvlocint=&hvlocphy=9042720&hvtargid=pla-524500659240. 12 Sept 2021.
49. Lopresti AL, Hood SD, Drummond PD. A review of lifestyle factors that contribute to important pathways associated with major depression: diet, sleep and exercise [Online]. J Affect Disord. 2013:12–27. http://researchrepository.murdoch.edu.auhttp://researchrepository.murdoch.edu.au/13504http://tweaket.com/CPGenerator/?id=13504of1
50. Parletta N, Zarnowiecki D, Cho J, Wilson A, Bogomolova S, Villani A, Itsiopoulos C, Niyonsenga T, Blunden S, Meyer B, Segal L, Baune BT, O'Dea K. A Mediterranean-style dietary intervention supplemented with fish oil improves diet quality and mental health in people with depression: a randomized controlled trial (HELFIMED). Nutr Neurosci. 2019;22:474–87. https://doi.org/10.1080/1028415X.2017.1411320.
51. Jacka FN, O'Neil A, Opie R, Itsiopoulos C, Cotton S, Mohebbi M, Castle D, Dash S, Mihalopoulos C, Chatterton M, Lou M, Brazionis L, Dean OM, Hodge AM, Berk M. A randomised controlled trial of dietary improvement for adults with major depression (the "SMILES" trial). BMC Med. 2017;15 https://doi.org/10.1186/s12916-017-0791-y.

52. Segal L, Twizeyemariya A, Zarnowiecki D, Niyonsenga T, Bogomolova S, Wilson A, O'Dea K, Parletta N. Cost effectiveness and cost-utility analysis of a group-based diet intervention for treating major depression–the HELFIMED trial. Nutr Neurosci. 2020;23:770–8. https://doi.org/10.1080/1028415X.2018.1556896.
53. Opie RS, O'Neil A, Itsiopoulos C, Jacka FN. The impact of whole-of-diet interventions on depression and anxiety: a systematic review of randomised controlled trials. Public Health Nutr. 2015;18:2074–93.
54. Yehuda S, Rabinovitz S, Mostofsky DI. Mixture of essential fatty acids lowers test anxiety. Nutr Neurosci. 2005;8:265–7. https://doi.org/10.1080/10284150500445795.
55. Buydens-Branchey L, Branchey M, Hibbeln JR. Associations between increases in plasma n-3 polyunsaturated fatty acids following supplementation and decreases in anger and anxiety in substance abusers. Prog Neuro-Psychopharmacol Biol Psychiatry. 2008;32:568–75. https://doi.org/10.1016/J.PNPBP.2007.10.020.
56. The leading voice in nutrition and dietetics | Dietitians Australia [Online]. 2020. https://dietitiansaustralia.org.au/. 15 Apr 2023.
57. Commission on Dietetic Registration [Online]. 2021. https://www.cdrnet.org/RDN. 15 Apr 2023.
58. What is Allied Health? ASAHP [Online]. 2020. https://www.asahp.org/what-is. 15 Apr 2023.
59. About Accredited Programs [Online]. [date unknown]. https://www.eatrightpro.org/acend/accredited-programs/about-accredited-programs. 15 Apr 2023.
60. Hager MH. Clinical privileging for registered dietitians: a regulatory perspective. J Am Diet Assoc. 2007;107:558–60. https://doi.org/10.1016/j.jada.2007.02.028.
61. Nutritionist vs Dietitian: Degree and Careers Comparison [Online]. 2023. https://www.publichealthdegrees.org/careers/become-registered-dietitian/dietitian-vs-nutritionist/. 15 Apr 2023.
62. Nutritionist | goarmy.com [Online]. [date unknown]. https://www.goarmy.com/careers-and-jobs/career-match/science-medicine/physical-mental-health/68m-nutrition-care-specialist.html. 15 Apr 2023.
63. Myers CM, Kim JS, Florian JP. (2018). Effects of repeated long-duration water immersions on skeletal muscle performance in well-trained male divers. Eur J Appl Physiol. 2018;118(10):2065–2076. https://doi.org/10.1007/s00421-018-3928-6
64. Myers CM, Kim J, Florian JP. Consecutive, long-duration hyperoxic immersions effect on skeletal muscle performance in well trained, male divers. In: American College of Sports Medicine national conference; 2019.
65. Myers CM, Kim J-S, Florian JP. Effects of consecutive, long- duration water immersions on skeletal muscle performance in well trained, male divers. J Am Coll Sports Med. 2017;49:S215.
66. Myers CM, Kim J-S, Musilli M, Mccully K, Florian JP. Effects of resting, consecutive, long-duration water immersions on neuromuscular endurance in well-trained males. Front Physiol. 2018;9:977. https://doi.org/10.3389/fphys.2018.00977.
67. Powers SK, Howley ET. Exercise physiology: theory and application to fitness and performance. 8th ed. New York City: McGraw-Hill Humanities/Social Sciences/Languages; 2011.
68. Costanzo L. Physiology. 5th ed. [Online]. http://www.amazon.com/Physiology-STUDENT-CONSULT-Online-Costanzo/dp/145570847X/ref=sr_1_1?s=books&ie=UTF8&qid=1431694542&sr=1-1&keywords=costanzo+physiology+5th+edition. 15 May 2015.
69. Costanzo LS. Physiology [Online]. 7th ed. https://www.vitalsource.com/products/costanzo-physiology-linda-costanzo-v9780323793346. 20 Mar 2022.
70. Betts JG, Desaix P, Johnson E (Edward W), Johnson JE, Korol O, Kruse D, Poe B, Wise J, Womble MD, Young KA, OpenStax College, open textbook library, OpenStax (nonprofit organization). Anatomy and physiology; 2021.
71. Jeukendrup AE. Nutrition for endurance sports: Marathon, triathlon, and road cycling. J Sports Sci. 2011;29 https://doi.org/10.1080/02640414.2011.610348.

Chapter 6
The Psychological and Cognitive Domains

Introduction

Human performance optimization is inherently intertwined with the Psychological and Cognitive Domains. The Psychological and Cognitive Domains support each other and service the whole HWS within the Healthcare Continuum (Fig. 6.1). By understanding and applying knowledge of the human mind and behavior, psychologists and cognitive psychologists can develop strategies that enhance physical performance and optimize cognitive function and psychological resilience. Navigating the complexities of human performance optimization necessitates a diverse team of professionals, each bringing unique expertise to the table. Among these, two roles stand out due to their focused understanding of the human mind and its limitless capacities: psychologists and cognitive psychologists. This chapter aims to introduce these vital professions, showcasing their significant contributions to human performance optimization programs.

Psychologists play an instrumental role in performance optimization programs, focusing on mental health, emotional resilience, stress management, and interpersonal communication. Their expertise is crucial in helping individuals and teams overcome mental roadblocks, cope with high-pressure situations, and build cognitive strategies for peak performance. Additionally, they are essential in identifying and addressing any underlying mental health issues that may impact performance.

Cognitive psychologists bring a nuanced understanding of cognitive processes like attention, memory, perception, and problem-solving to the mix. Their role in performance optimization involves improving these mental skills, thereby enhancing decision-making, learning efficiency, and overall cognitive performance. This includes designing and implementing cognitive training protocols, interpreting cognitive assessment results, and utilizing technology like mental simulations to provide tailored interventions.

C. Myers, *The Human Weapon System*,
https://doi.org/10.1007/978-3-031-45061-7_6

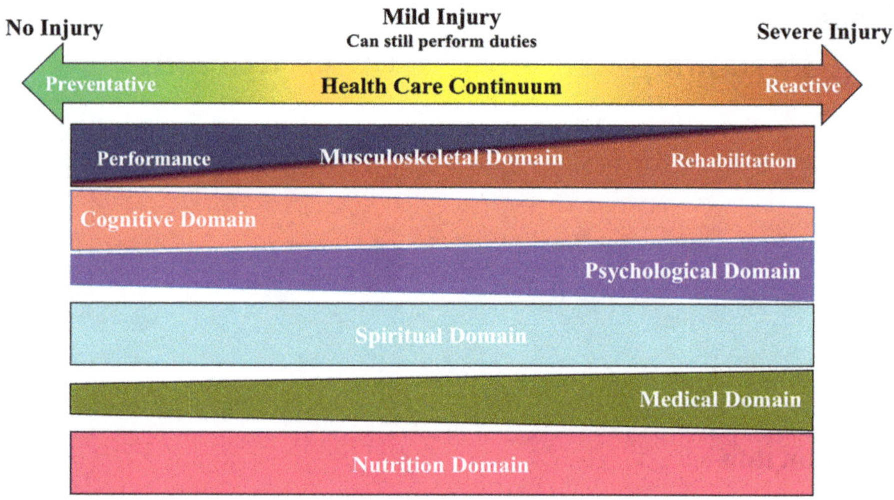

Fig. 6.1 The psychological and cognitive domains support each other and service the whole HWS within the Healthcare Continuum (same as Fig. 3.7)

This chapter delves deeper into the roles and contributions of these two professions. It will explore their training, methodologies, and the specific strategies they use to foster performance enhancement. Furthermore, this chapter illustrates how they work synergistically with other professionals in a performance optimization program. They use their specialized knowledge to contribute to the holistic betterment of an individual's performance.

The Psychological Domain

The Psychological Domain is primarily comprised of a psychologist. A licensed psychologist is a privileged clinician with a doctoral degree in psychology. The license is obtained within the state where the psychologist is practicing. However, in the military, the psychologist only needs to be licensed in a single state [1–3].

Depending on the size and scope of the HPO program, the domain may contain an additional member. The licensed clinical social worker (LCSW) can work independently with the psychologist. Smaller programs may only have an LCSW. An LCSW must have a master's degree in social work, pass an internship, and be licensed in the state of practice. Like psychologists, LCSWs are privileged practitioners [4].

Education and Licensure

All psychologists must obtain a bachelor's degree. Many choose to pursue a degree in Psychology, but the degree does not need to be so. At a minimum, aspiring psychologists must be introduced to various aspects of psychology, such as developmental, abnormal, social, cognitive, and personality psychology. This undergraduate education lays the foundation for understanding human behavior, mental processes, research methods, and statistics [36].

Following undergraduate studies, the next step is a graduate degree, typically a doctoral degree (Ph.D. or Psy.D.) in Psychology. Ph.D. is research focused, involving completing a dissertation and extensive training in research methods, statistics, and teaching. Psy.D. is clinically focused, providing training in psychological testing, diagnosis, treatment planning, and the provision of therapy [36]. Both degrees usually require supervised clinical internships. This stage typically takes 4–6 years, depending on the degree program. It is important to note that some psychologists choose to earn a master's degree before proceeding to the doctoral level, although this is not a requirement for all postgraduate programs [36].

A practicing psychologist must acquire a license. Licensing requirements vary from state to state or country to country. However, the licensure requirements often include the following:

- A doctoral degree from an accredited institution.
- A specific number of supervised clinical experience hours.
- Passing a licensure exam.

The Examination for Professional Practice in Psychology (EPPP) is the standard licensing exam in the United States [37]. This exam covers many areas, including biological bases of behavior, cognitive-affective bases of behavior, social and multicultural bases of behavior, growth, and life span growth and development; assessment, diagnosis, and treatment of behavioral and mental health diseases; and research methods and statistics [37].

The education and licensure requirements for an LCSW differ from those of a psychologist. Similarly, an LCSW begins by obtaining a bachelor's degree. While many pursue a Bachelor's in Social Work (BSW), some might major in Psychology, Sociology, or Human Services. This undergraduate program introduces students to the fundamentals of social work and typically includes courses on human behavior, social welfare policies, diversity, and the basics of research methods.

After earning a bachelor's degree, the aspiring LCSW must earn a Master's Degree in Social Work (MSW) from a Council on Social Work Education (CSWE) accredited program [1, 2]. This typically takes about 2 years. The curriculum includes classroom education and practical fieldwork experience [1]. Courses often cover advanced topics like clinical practice, social policy, human behavior and the social environment, research methods, and social work practice with individuals, families, and groups [1].

After obtaining the MSW, most jurisdictions require a period of supervised clinical experience before licensure can be achieved. This typically consists of 2 years or 3000 hours of supervised clinical social work experience [3, 4]. An LCSW or a similar professional should conduct the supervision, and it allows the social worker to apply their skills in a practical setting while receiving feedback and guidance.

Becoming an LCSW requires passing the Clinical Level Association of Social Work Boards (ASWB) examination for the United States [5]. This exam covers various topics, including human growth and development, concepts of abuse and neglect, assessment and intervention planning, assessment of methods and techniques, intervention processes, and values and ethics [6, 7].

Responsibilities

The roles and responsibilities of a psychologist and an LCSW are vastly different in an HPO program. Mainly, these differences are due to two items. The first is due to the outcomes of the needs analysis. As laid out in Chap. 9, the needs analysis determines the focus and scope of the HPO program. The second aspect is the training differences between a psychologist and an LCSW. This section details the roles and responsibilities of the psychologist and LCSW in an HPO program.

In HPO programs, psychologists apply the principles of psychology to improve individuals' performance, resilience, and overall well-being, often within organizational settings. They use various psychometric tools and techniques to assess individual capabilities, cognitive functions, personality traits, and emotional states that may impact performance. Then based on assessments, psychologists design and implement interventions to enhance individual and team performance. Interventions could include cognitive training, stress management techniques, individual and group therapy/counseling sessions, or coaching for improved communication and leadership skills [8, 9]. Psychologists often provide performance coaching, leveraging psychological insights to help individuals improve their productivity, manage stress, enhance motivation, and cultivate resilience [8, 9].

The roles of LCSWs in HPO programs are vastly different from those of a psychologist. LCSWs focus on the social and mental health issues of the HWS in the organizational setting. These specialists perform this role through individual and group therapy sessions, advocacy, and case management [10].

Both psychologists and LCSWs bring unique skills and perspectives to HPO programs, and their roles often intersect. Together, they form a multidisciplinary team that addresses the multifaceted nature of human performance. They prioritize evidence-based practice and work collaboratively with individuals to enhance their performance and well-being.

The Cognitive Domain

Education and Licensure

The educational path to becoming a cognitive psychologist is very similar to that of a psychologist. Aspiring cognitive psychologists typically obtain a Bachelor's Degree in Psychology or a closely related field. Undergraduate coursework provides a broad understanding of psychological theories, research methods, statistics, and foundational knowledge in various subfields of psychology [1–3]. In some cases, students may have the opportunity to specialize in cognitive psychology by selecting relevant elective courses, such as cognitive processes, perception, attention, memory, and learning [1–3].

Graduate education is where cognitive psychologists' and psychologists' education paths diverge. Cognitive psychologists obtain doctoral degrees in cognitive psychology or a related field from an American Psychological Association (APA)-accredited program. Doctoral programs, such as a Ph.D. or Psy.D. in Psychology, provide in-depth training and research opportunities in cognitive psychology. During graduate studies, aspiring cognitive psychologists focus on advanced coursework that delves into cognitive neuroscience, cognitive assessment, cognitive development, cognitive psychology theories, and experimental research methods. Furthermore, these programs involve conducting original research in cognitive psychology, culminating in a dissertation that contributes to the field's knowledge base [1, 2].

As part of their doctoral training, aspiring cognitive psychologists typically engage in supervised practicum experiences and internships. These placements allow students to gain practical skills in administering cognitive assessments, designing experiments, analyzing data, and conducting research under the guidance of experienced professionals. The hands-on programs provide valuable training in applying cognitive psychology principles to real-world settings, such as clinics, research laboratories, or academic institutions.

Licensure

After completing a doctoral program, cognitive psychologists must meet specific licensure requirements to practice independently. Licensure requirements vary by jurisdiction but often involve completing the supervised postdoctoral experience, passing the EPPP and the Professional Standards exams, and specialty-specific requirements.

Cognitive psychologists' educational and licensure requirements reflect the specialized knowledge and skills necessary to understand and study human cognition. Through undergraduate and graduate education, supervised training, licensure, and continued professional development, cognitive psychologists acquire the expertise to conduct research, assess cognitive abilities, and apply cognitive principles to optimize human performance and well-being.

Responsibilities

Cognitive psychologists play a vital role in human performance optimization programs, applying their expertise in understanding human cognition, behavior, and mental processes to enhance individual and team performance. These professionals deeply understand how humans perceive, process, and interpret information. They study cognitive functions such as attention, memory, decision-making, problem-solving, and learning. Within a human performance optimization program, cognitive psychologists apply their knowledge to assess the mental demands of specific tasks, identify cognitive strengths and limitations, and develop strategies to optimize cognitive performance.

Cognitive psychologists design and administer assessments to evaluate cognitive abilities and performance. These assessments can include tests, questionnaires, and cognitive tasks to measure attention, memory, problem-solving, and other cognitive functions. By conducting cognitive assessments, psychologists can identify areas of strength and weakness in individuals or teams, highlighting specific cognitive skills that may need improvement or optimization.

These professionals develop and implement cognitive training programs to enhance cognitive skills and optimize performance. These programs may include strategies to improve attention, memory, decision-making, and problem-solving abilities. Psychologists work with individuals or teams through targeted interventions to enhance their cognitive skills and overcome mental limitations that may hinder performance. This approach can involve attentional control training, memory strategies, cognitive-behavioral therapy, or mindfulness practices.

Finally, cognitive psychologists collaborate with other professionals, such as physicians, physiotherapists, nutritionists, and coaches, to promote holistic human performance optimization. They contribute their expertise to improve team dynamics, communication, collaboration, and decision-making. These professionals analyze the impact of cognitive factors on teamwork, identifying potential barriers and developing strategies to enhance cognitive synergy and team performance.

By integrating cognitive psychology principles and techniques, cognitive psychologists contribute to optimizing human performance in various domains. Their understanding of human cognition, combined with their expertise in assessment, training, and intervention, helps individuals and teams unlock their cognitive potential, overcome limitations, and achieve peak performance.

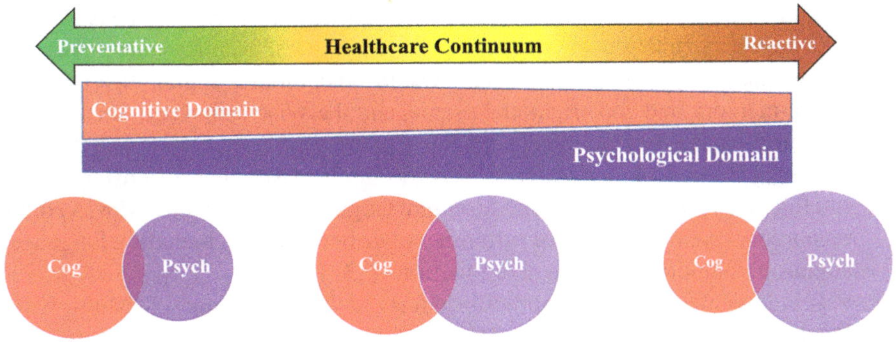

Fig. 6.2 The cognitive and psychological domains' relationship concerning the Healthcare Continuum

Interlacing the Cognitive and Psychological Domains

As stated earlier, the composition of these two domains is determined by the needs analysis and focus and scope of the HPO program. Furthermore, available funding will evaluate the ability to have both domains. Having the Cognitive and Psychological Domains completes the ability to address the HWS's mental and cognitive needs fully.

The Cognitive and Psychological Domains work together within an HPO program. Even though they are separate domains, they work together similarly to what an exercise physiologist (EP) and a physical therapist (PT) do in the MSK Domain. As shown in Fig. 6.1, an HWS interaction with each domain depends on where the HWS falls within the Healthcare Continuum. When an HWS requires emotional or mental health intervention (far right side of the Healthcare Spectrum), the primary interaction is with the Psychological Domain team. However, when the HWS is more focused on performance and sustaining resiliency (far left side of the Healthcare Spectrum), the HWS tends to interact more with the Cognitive Domain team (Fig. 6.2).

Performance Metrics of the Psychological and Cognitive Domains

Cognitive Domain

A cognitive psychologist may use various performance metrics within an HPO program. Such metrics often measure cognitive functions like attention, memory, decision-making, reaction time, and problem-solving abilities. Here are some examples:

1. Cognitive test scores: these could include standardized tests that measure specific cognitive functions. For instance, the Stroop Test evaluates cognitive flexibility and processing speed, the Rey-Osterrieth Complex Figure Test assesses visual memory and constructional praxis, and the Wisconsin Card Sorting Test measures executive functions [11, 12].
2. Neuropsychological assessment scores: tools like the Wechsler Adult Intelligence Scale (WAIS) or the Montreal Cognitive Assessment (MoCA) offer comprehensive evaluations of different cognitive domains, including language, attention, memory, and executive functions [13, 14].
3. Psychological test scores: standardized tests such as the Minnesota Multiphasic Personality Inventory (MMPI) or the Millon Clinical Multiaxial Inventory (MCMI) can be used to identify personal, social, and coping characteristics that could influence performance [15, 16].
4. Emotional intelligence measures: tools like the Mayer-Salovey-Caruso Emotional Intelligence Test (MSCEIT) or Emotional Quotient Inventory (EQ-i) measure emotional intelligence linked to performance in many contexts [17, 18].
5. Self-efficacy measures: this refers to an individual's ability to believe that they can accomplish any task. Tools like the General Self-Efficacy Scale (GSE) can be used to measure this trait [19].
6. Motivation measures: surveys or questionnaires, such as the Motivated Strategies for Learning Questionnaire (MSLQ), can assess an individual's drive to achieve their goals [20].
7. Stress and resilience measures: psychologists often use instruments such as the Perceived Stress Scale (PSS) or the Connor-Davidson Resilience Scale (CD-RISC) to measure an individual's perceived stress levels and resilience to stress [21, 22].
8. Workplace performance metrics: these may include key performance indicators (KPIs) related to the individual's role, 360-degree feedback from colleagues and supervisors, and task completion and productivity measures.
9. Personality measures: tools like the Myers-Briggs Type Indicator (MBTI) or the Big Five Personality Test can provide insights into an individual's personality traits, influencing their approach to work and interaction with others [23, 24].
10. Mindfulness and attention measures: instruments like the Mindful Attention Awareness Scale (MAAS) can measure an individual's mindfulness linked to improved performance and stress management [25, 26].
11. Job satisfaction and engagement measures: measures like the Utrecht Work Engagement Scale (UWES) or the Job Satisfaction Survey (JSS) can provide insights into how engaged and satisfied an individual is with their job, which is often linked to their performance [27].
12. Physical health measures: given the interaction between physical and mental health, measures such as sleep quality, physical activity levels, or even measures of physiological functioning (e.g., heart rate variability) could be helpful [28].

13. Reaction time: this measure helps assess cognitive processing speed, which can be crucial in high-stress or time-sensitive situations [12].
14. Working memory capacity: tools like Digit Span or N-back tasks can be used to evaluate the capacity of working memory, which plays a crucial role in comprehension, learning, and reasoning [12].
15. Attention span: psychologists often use Continuous Performance Tests (CPT) to measure a person's sustained and selective attention span [29, 30].
16. Problem-solving abilities: tasks like the Tower of London or Raven's Progressive Matrices can measure a person's problem-solving and reasoning skills [31, 32].
17. Decision-making abilities: the Iowa Gambling Task or similar tests can be used to assess decision-making skills, particularly those involving risk and reward evaluation.
18. Mental flexibility: task-switching or set-shifting tasks (e.g., the Trail Making Test) can evaluate cognitive flexibility and executive functioning.
19. Stress tolerance: measures of physiological responses to stress, such as heart rate variability or cortisol levels, can provide insights into a person's ability to handle stress.
20. Sleep quality: given the strong relationship between sleep and cognitive function, tools like actigraphy or self-report sleep diaries can provide essential sleep quality metrics and their impact on performance [28].
21. Self-report measures: questionnaires or surveys about perceived cognitive function, stress, fatigue, and other relevant factors can provide valuable subjective data to supplement objective metrics.

Psychological Domain

Similarly, a psychologist might use a variety of performance metrics that measure not only cognitive abilities but also emotional and behavioral aspects that are crucial for optimal performance. Here are some potential metrics:

1. Psychological test scores: standardized tests such as the Minnesota Multiphasic Personality Inventory (MMPI) or the Millon Clinical Multiaxial Inventory (MCMI) can identify personal, social, and coping characteristics that could influence performance [33].
2. Emotional intelligence measures: tools like the Mayer-Salovey-Caruso Emotional Intelligence Test (MSCEIT) or Emotional Quotient Inventory (EQ-i) measure emotional intelligence linked to performance in many contexts [11, 12, 34].
3. Self-efficacy measures: this refers to an individual's ability to believe he can accomplish any task. Tools like the General Self-Efficacy Scale (GSE) can be used to measure this trait [35].

4. Motivation measures: surveys or questionnaires, such as the Motivated Strategies for Learning Questionnaire (MSLQ), can assess an individual's drive to achieve their goals [20].
5. Stress and resilience measures: psychologists often use instruments such as the Perceived Stress Scale (PSS) or the Connor-Davidson Resilience Scale (CD-RISC) to measure an individual's perceived stress levels and resilience to stress [21, 22].
6. Workplace performance metrics: these may include key performance indicators (KPIs) related to the individual's role, 360-degree feedback from colleagues and supervisors, and task completion and productivity measures.
7. Personality measures: tools like the Myers-Briggs Type Indicator (MBTI) or the Big Five Personality Test can provide insights into an individual's personality traits, influencing their approach to work and interaction with others [23, 24].
8. Mindfulness and attention measures: instruments like the Mindful Attention Awareness Scale (MAAS) can measure an individual's mindfulness linked to improved performance and stress management [25, 26].
9. Job satisfaction and engagement measures: measures like the Utrecht Work Engagement Scale (UWES) or the Job Satisfaction Survey (JSS) can provide insights into how engaged and satisfied an individual is with their job, which is often linked to their performance [27].
10. Physical health measures: given the interaction between physical and mental health, measures such as sleep quality, physical activity levels, or even measures of physiological functioning (e.g., heart rate variability) could be helpful [28].

As with cognitive measures, the specific metrics used in each HPO program would depend on the goals of that program and the roles and responsibilities of the individuals involved.

Conclusion

The Cognitive and Psychological Domains focus on the mental and cognitive health and abilities of the HWS. The education and licensure requirements for these health professionals are very stringent. The standards exist to ensure the HWS receives the best treatment and training possible. Additionally, many HPO programs may only have one of the two domains. This aspect is determined by the needs analysis and available funding. When both domains exist in the same HPO program, the HWS can access the full spectrum of cognitive and psychological training and treatment.

References

1. Value and benefits of CSWE accreditation [Online]. [date unknown]. https://www.cswe.org/accreditation/about/value-benefits/. 16 July 2023.
2. West Virginia Board of Social Work Licensure [Online]. [date unknown]. https://www.wvsocialworkboard.org/Licensure/License-Types/Regular-License/Level-C-LCSW. 16 July 2023.
3. Policy on 3000 hours of clinical experience [Online]. https://www.tn.gov/content/dam/tn/health/documents/Social_Work_Board_Policy_on_3000_Hours_of_Clinical_Experience.pdf. 16 July 2023.
4. LCSW breakdown of required experience [Online]. https://www.bbs.ca.gov/pdf/forms/lcs/lcs_exp_chart.pdf. 16 July 2023.
5. Licenses – Association of Social Work Boards [Online]. 2023. https://www.aswb.org/licenses/. 16 July 2023.
6. ASWB masters of social work license exam outlines and KSAs [Online]. https://www.aswb.org/wp-content/uploads/2020/12/2018-Masters.pdf. 16 July 2023.
7. Exam – Association of Social Work Boards [Online]. 2023. https://www.aswb.org/exam/. 16 July 2023.
8. Psychologists: occupational outlook handbook. U.S. Bureau of Labor Statistics [Online]. 2022. https://www.bls.gov/ooh/life-physical-and-social-science/psychologists.htm#. 16 July 2023.
9. Wahass SH. The role of psychologists in health care delivery. [Online]. J Fam Community Med 12: 63–70, 2005. http://www.ncbi.nlm.nih.gov/pubmed/23012077. 16 July 2023.
10. Career guide: licensed clinical social worker (LCSW) [Online]. [date unknown]. https://www.publichealthonline.org/lcsw-licensed-clinical-social-worker/. 16 July 2023.
11. Kaplan RM. In: Robert M, Saccuzzo DP, editors. Psychological testing: principles, applications, & issues. New Orleans: White Press Academic; 2018.
12. Cohen RJ, Tobin RM, Schneider WJ. Psychological testing and assessment: an introduction to tests and measurement. New York: McGraw-Hill; 2022.
13. Wechsler D. Wechsler adult intelligence scale, WAIS–IV. 4th ed. San Antonio: NCS Pearson, 22; 2008.
14. Nasreddine ZS, Phillips NA, Bédirian V, Charbonneau S, Whitehead V, Collin I, Cummings JL, Chertkow H. The Montreal cognitive assessment, MoCA: a brief screening tool for mild cognitive impairment. J Am Geriatr Soc. 2005;53 https://doi.org/10.1111/j.1532-5415.2005.53221.x.
15. Mohammadi MR, Delavar A, Hooshyari Z, Shakiba A, Salmanian M, Ghandi F, Ahmadi A, Farnoody N. Psychometric properties of the Persian version of Millon Clinical Multiaxial Inventory-IV (MCMI-IV). Iran J Psychiatry. 2021;16 https://doi.org/10.18502/ijps.v16i1.5378.
16. Pona AA, Marek RJ, Panigrahi E, Ben-Porath YS. Examination of the reliability and validity of the Minnesota Multiphasic Personality Inventory-3 (MMPI-3) in a preoperative bariatric surgery sample. https://doi.org/10.1007/s10880-022-09908-2
17. Dawda D, Hart SD. Assessing emotional intelligence: reliability and validity of the bar-on emotional quotient inventory (EQ-i) in university students. Pers Individ Dif. 2000;28 https://doi.org/10.1016/S0191-8869(99)00139-7.
18. Mayer J, Saloyev P, Caruso D. The Mayer-Salovey-Caruso Emotional Intelligence Test (MSCEIT). North Tonawanda: Multi Health Systems; 2002.
19. Schwarzer R, Jerusalem M. Generalized Self-Efficacy scale. In J. Weinman, S. Wright, & M. Johnston, Measures in health psychology: A user's portfolio. Causal and control beliefs (pp. 35–37). Windsor, UK: NFERNELSON; 1995.
20. Jackson CR. Validating and adapting the motivated strategies for learning questionnaire (MSLQ) for STEM courses at an HBCU. AERA Open. 2018;4 https://doi.org/10.1177/2332858418809346.
21. Connor KM, Davidson JRT. Development of a new resilience scale: the Connor-Davidson resilience scale (CD-RISC). Depress Anxiety. 2003;18 https://doi.org/10.1002/da.10113.

22. Townsend S, Medvedev ON. Perceived stress scale (PSS). In: Handbook of assessment in mindfulness research. Cham: Springer; 2022.
23. Choong EJ, Varathan KD. Predicting judging-perceiving of Myers-Briggs Type Indicator (MBTI) in online social forum. PeerJ. 2021;9:e11382. https://doi.org/10.7717/peerj.11382.
24. Boyle GJ. Myers-Briggs type indicator (MBTI): some psychometric limitations. Aust Psychol. 1995;30 https://doi.org/10.1111/j.1742-9544.1995.tb01750.x.
25. Reyes-Bossio M, Zapparigli EL, Caycho-Rodríguez T, Carbajal-León C, Castaman LAO, Pino GLH, Sanchez RC, Barbosa-Granados S. Cross-cultural validity of the five items Mindful Attention Awareness Scale (MAAS-5) in Peru and Mexico during the COVID-19 pandemic. Psicologia: Reflexao e Critica. 2022;35 https://doi.org/10.1186/s41155-022-00218-y.
26. MacKillop J, Anderson EJ. Further psychometric validation of the mindful attention awareness scale (MAAS). J Psychopathol Behav Assess. 2007;29:289–93. https://doi.org/10.1007/s10862-007-9045-1.
27. Schaufeli WB, Bakker AB, Salanova M. The measurement of work engagement with a short questionnaire: a cross-national study. Educ Psychol Meas. 2006;66 https://doi.org/10.1177/0013164405282471.
28. Powers SK, Howley ET. Exercise physiology: theory and application to fitness and performance. 8th ed. New York City: McGraw-Hill Humanities/Social Sciences/Languages; 2011.
29. Elbaum T, Braw Y, Lev A, Rassovsky Y. Attention-deficit/hyperactivity disorder (Adhd): integrating the MOXO-dCPT with an eye tracker enhances diagnostic precision. Sensors (Switzerland). 2020;20 https://doi.org/10.3390/s20216386.
30. Winter D, Braw Y. Online search strategies utilized in feigning attention deficit/hyperactivity disorder (ADHD) while performing a continuous performance test (CPT). https://doi.org/10.1080/23279095.2022.2128356.
31. Brouwers SA, Van de Vijver FJR, Van Hemert DA. Variation in Raven's progressive matrices scores across time and place. Learn Individ Differ. 2009;19 https://doi.org/10.1016/j.lindif.2008.10.006.
32. Song JHH, Loyal S, Lond B. Metacognitive awareness scale, domain specific (MCAS-DS): assessing metacognitive awareness during Raven's progressive matrices. Front Psychol. 2021;11 https://doi.org/10.3389/fpsyg.2020.607577.
33. Craig RJ. The Millon clinical multiaxial inventory: a clinical research information synthesis. Mahwah: L. Erlbaum Associates; 1993.
34. Emmerling R. Mayer-Salovey-Caruso Emotional Intelligence Test (MSCEIT) [Online]. 2022. https://www.eiconsortium.org/measures/msceit.html. 19 July 2023.
35. Chen G, Gully SM, Eden D. Validation of a new general self-efficacy scale. Organ Res Methods. 2001;4:62–83. https://doi.org/10.1177/109442810141004.
36. Chamlou N. How to become a clinical psychologist; 2023. https://www.psychology.org/careers/clinical-psychologist/
37. Dittmann M. What you need to know to get licensed; 2004. Retrieved March 14, 2023, from https://www.apa.org/gradpsych/2004/01/get-licensed

Chapter 7
Medical Domain

Introduction

In human performance optimization, the medical domain ensures individuals' well-being and peak performance. Integrating highly skilled healthcare professionals, such as physician assistants (PAs) and medics, within performance optimization programs has proven instrumental in achieving optimal results. This chapter explores the vital contributions of PAs and medics in enhancing human performance and examines the collaborative efforts that drive success in this domain.

In today's fast-paced world, where demands on HWS' physical and mental capacities are constantly increasing, developing comprehensive strategies to optimize human performance has become essential. Whether elite athletes striving for record-breaking achievements or military personnel operating in high-stress environments, the need for a holistic approach to health and performance has never been more critical.

With their extensive medical knowledge and practical skills, physician assistants have emerged as invaluable assets in optimizing human performance. Working closely with other healthcare professionals, PAs contribute to preventing, diagnosing, and treating injuries and illnesses that can impact an individual's performance. Their ability to provide primary healthcare, administer medications, and perform minor medical procedures makes them an integral part of the multidisciplinary teams working toward performance enhancement.

Similarly, medics often have a background in emergency medical services or military medicine, bringing unique expertise to human performance optimization. These highly trained individuals can swiftly assess and address acute injuries or medical emergencies, ensuring rapid and effective interventions. Their competence in providing immediate care in demanding situations is vital to maintaining the health and performance of individuals operating in high-risk environments.

C. Myers, *The Human Weapon System*,
https://doi.org/10.1007/978-3-031-45061-7_7

The collaborative efforts of PAs and medics within human performance optimization programs foster an environment of comprehensive care. By working together, they combine their specialized knowledge and skills, allowing for a broader and more thorough assessment of an individual's health status and becoming more preventative versus reactive in medical treatments.

Medical Domain Team

The medical domain team can be comprised of any medical professional. The needs analysis, funding available, and HPO program focus determine the members of this domain. The most common medical professions found in the medical domain of an HPO program are physician assistants (PA) and medics/emergency medical technicians (EMTs).

Physician Assistant (PA)

A PA is crucial in a human performance optimization program. They possess the qualifications and responsibilities necessary to contribute effectively to such a program. Here's a summary of their capabilities and responsibilities.

Education and Qualifications

Qualified PAs with extensive medical education and training typically culminate with a master's degree and clinical rotations. PAs must pass a national certification exam upon completing their education and training.

1. Education

Typically, PAs have a bachelor's degree. Some higher-level PA programs require specialized courses but typically not specialized majors [1]. However, many aspiring PAs will obtain degrees in premedical or health-related fields, such as biology, chemistry, or healthcare administration [1]. If an aspiring PA does not pursue one of the specialized degrees, advanced PA programs require specialized classes in natural sciences, including biology, chemistry, physics, and anatomy. Additional courses in psychology, sociology, and medical ethics, which can provide a broader understanding of patient care and healthcare, may be required. These requirements depend on the graduate-level PA program the aspiring PA wishes to attend.

A PA master's degree has several names. These programs may be referred to as Master of Physician Assistant Studies, Master of Health Sciences (MHS), or Master of Clinical Health Services (MCHS) [2]. No matter what the program's name is, these programs are 2–3 years long. PA graduate programs integrate classroom instruction with hands-on clinical experiences. Aspiring PAs learn critical medical knowledge such as diagnostic and treatment skills and patient care techniques. Graduate PA curriculums also cover pharmacology, pathophysiology, clinical medicine, and medical ethics [2].

2. Clinical Rotations

Clinical rotations form a vital component of a graduate-level PA program. These rotations provide hands-on experience and practical expertise. Aspiring PAs develop critical thinking and clinical skills necessary for diagnosis and make evidence-based decisions during each rotation. Furthermore, PAs are exposed to several medical specialties, allowing for a well-rounded curriculum.

The structure and length of the clinical rotations can vary from program to program but have a core set of rotations. Typically, a PA will have over 2000 h of supervised clinical rotations by graduation. The rotations occur in clinics, hospitals, outpatient settings, etc., all under the supervision of physicians or experienced PAs. The supervising preceptor structures each rotation. This means variations can occur from preceptor to preceptor even though aspiring PAs are conducting similar subject matter clinical rotations. However, clinical rotations have a set of core rotations. These rotations ensure all PAs are proficient in critical medical specialties, such as family medicine, internal medicine, surgery, emergency medicine, pediatrics, obstetrics, and gynecology.

During each rotation, PAs learn several hard and soft skills. Some of the PA's hard skills are performing patient assessments, collecting and interpreting diagnostic tests, creating treatment plans, and collaborating with a team. Furthermore, PAs are exposed to different patient populations, diseases, healthcare administration, and other specialized medical practices. The other side of clinical rotations is soft skills. These are skills that allow a professional to interact well with a team. Some of these skills are professionalism, ethical conduct, and communication. These skills are necessary to foster inter- and intraprofessional relationships. Ultimately, these hard and soft skills promote teamwork and enhance patient care.

3. Certification

Certification is a crucial step in the professional journey of a physician assistant (PA). This section explores the certification requirements and process, emphasizing the importance of certification in ensuring competency, maintaining professional standards, and enabling PAs to provide safe and effective patient care.

PAs are certified through the National Commission on Certification of Physician Assistants (NCCPA) . This organization is the only certifying organization in the United States. Similar organizations exist in other countries. The NCCP sets the certification standards. Additionally, the organization administers the Physician Assistant National Certifying Examination (PANCE) [3].

The PANCE is a comprehensive examination that assesses the knowledge and skills necessary to become a PA. This examination covers various medical topics and clinical scenarios typically seen by PAs in the field. Additionally, the PANCE assesses critical thinking abilities, clinical reasoning skills, and the application of medical knowledge to patient care in internal medicine, surgery, pediatrics, obstetrics and gynecology, psychiatry, and more. Upon passing the examination, all PAs are issued the Physician Assistant-Certified (PA-C) credential. Finally, PAs must be licensed in the state they are practicing; each state has different requirements. For the US military, PAs need to be licensed in a single state and not necessarily in the state he is located while on active military orders [3].

Responsibilities

A PA has a myriad of responsibilities within an HPO program. The following is a brief list of those responsibilities:

1. Patient assessment: PAs are skilled at conducting comprehensive patient assessments, which include taking medical histories, performing physical exams, and ordering and interpreting diagnostic tests [4, 5].
2. Diagnosis and treatment: PAs collaborate with supervising physicians to diagnose medical conditions and develop treatment plans. They may prescribe medications, provide therapies, and perform minor procedures under physician supervision [4].
3. Preventive care: PAs emphasize preventive measures, such as health screenings, immunizations, and lifestyle counseling, to optimize patients' overall well-being and performance [4, 5].
4. Patient education: PAs educate patients about their conditions, treatment options, and the ways to enhance their performance. They offer guidance on managing chronic illnesses, adopting healthy habits, and understanding the importance of adherence to medical regimens [4].
5. Team collaboration: PAs work closely with a multidisciplinary team, including physicians, nurses, therapists, and other healthcare professionals, to ensure comprehensive care and optimize human performance [4, 5].
6. Continuity of care: PAs contribute to ongoing patient care by monitoring treatment progress, evaluating outcomes, and making necessary adjustments to treatment plans. They also facilitate referrals to specialists when needed [4, 5].
7. Research and evidence-based practice: PAs stay updated with medical advancements and evidence-based practices to deliver high-quality care. They may also participate in research and quality improvement initiatives [4].

As part of a human performance optimization program, a physician assistant brings a diverse skill set, extensive medical knowledge, and a commitment to patient care. They collaborate with healthcare teams, diagnose and treat medical conditions, promote preventive care, educate patients, and ensure continuity of care. Their contributions are vital in optimizing human performance, improving access to healthcare, and enhancing overall HWS well-being.

Medic

A medic in the military serves a critical role in providing medical care and support to service members. Here is a summary of their qualifications and responsibilities, along with how they can assist a physician assistant (PA) as part of a human performance optimization program:

Qualifications

1. Training

This section briefly describes the primary training all medics obtain regardless of military branch.

Military combat medics begin their training with initial entry training (IET). This course varies in length based on the service branch. Each IET prepares the medic for the physical, emotional, and psychological demands of military service. This training creates a baseline that every medic has, like other professions within their branch of service.

Upon the completion of IET, medics attend secondary training. The Army, the Marines, and, unofficially, the Navy name this training Advanced Individual Training (AIT), and the Air Force terms are Air Force Specialty Code (AFSC) training. This secondary training is where the medics build their foundation in basic medical knowledge and skills. This initial phase may include classroom instruction and hands-on training in anatomy, physiology, emergency medicine, trauma management, and essential life support. As training progresses, they learn how to assess and stabilize patients, administer first aid, control bleeding, manage fractures, and provide initial care in combat or emergencies. They develop a solid understanding of medical equipment, supplies, and protocols specific to the military healthcare system.

Also, during secondary training, medics further hone their medical skills in a combat-specific environment. This stringent and specialized training environment prepares the medics for the stresses and uniqueness of battlefield trauma. Medics learn how to triage in a stressful environment. This skill is critical to maximizing safe-saving strategies and supplies in a dynamic operational environment. Additionally, medics know specialized training, including tactical combat casualty care (TCCC), advanced trauma life support, advanced airway management, combat surgical techniques, prolonged field care, or the provision of care in chemical, biological, radiological, and nuclear (CBRN) environments. Medics may also receive training in operational medicine, which involves understanding the unique medical challenges faced by military personnel, such as environmental injuries, infectious diseases, and mental health concerns.

Responsibilities

Comparable to those listed for a PA, a medic has several areas of responsibility. The following explains just a tiny portion of these responsibilities.

1. Emergency care: medics are trained to provide immediate medical care in high-stress situations, including battlefield trauma, combat injuries, and medical emergencies. They stabilize patients, administer first aid, and initiate treatment until further medical support is available [4].
2. Triage and evacuation: medics assess and prioritize injured personnel during combat or disaster situations, triaging patients based on the severity of their conditions. They coordinate medical evacuation and ensure timely transportation to higher-level medical facilities [4].
3. Primary care: in noncombat situations, medics deliver primary healthcare to military personnel, including routine checkups, vaccinations, and minor illness or injury management. They assist with health screenings and monitor service members' overall health and readiness [4].
4. Support to physician assistants: medics work closely with PAs, assisting them in medical assessments, performing preliminary exams, gathering patient histories, and obtaining vital signs. They provide valuable support during procedures and treatment interventions and help with patient education (4).
5. Training and education: medics contribute to training programs, instructing military personnel on basic medical skills, first aid techniques, and injury prevention. They ensure that HWS is prepared to handle medical emergencies via TCCC training.
6. Field medical support: medics accompany military units on deployments or field exercises, offering immediate medical support and enhancing the unit's operational effectiveness. They contribute to maintaining the health and performance of service members in challenging environments [4].

Medics in the military bring specialized training and experience to provide medical care and support to service members. They assist PAs by working alongside them in various healthcare tasks, supporting medical assessments, providing primary care, and aiding in procedures. Medics are crucial in ensuring the well-being and optimization of human performance in military settings.

The medical domain composition may differ in the HWS professions of law enforcement and firefighters. The domain may be staffed with other medical specialties such as nurse practitioners (NPs), nurses, and physicians. Depending on the size, scope, and focus of the HPO program, the medical domain may have one or multiple medical professionals. This chapter will cover only some of these differing specialties' education and certification requirements. However, one difference that needs to be discussed is the difference between medics and emergency medical technicians (EMTs).

Medics and Emergency Medical Technicians (EMTs)

The medic and EMT roles are similar but significantly different. Both are dedicated to emergency care and life-saving interventions; however, their operational environments, training, and scope of practice distinguish them from each other.

One of the most apparent differences between combat medics and EMTs is the environment in which they operate. Combat medics, as suggested by their title, work in military settings, often directly on the battlefield. They aim to provide immediate medical care in austere and often hostile environments, stabilizing and evacuating casualties under fire. They must perform their duties under extraordinary stress, facing threats like gunfire, explosions, and potential chemical or biological agents.

On the other hand, EMTs work in civilian contexts, which can range from urban streets to rural highways or homes. They are tasked with responding to emergency medical calls, assessing patient conditions, and providing immediate treatment for critical injuries or illnesses. While EMTs often operate in high-stress situations like motor vehicle accidents, fires, or natural disasters, they typically do not face the same potential hostility or tactical complexity inherent in a combat zone.

The training of these two medical specialties is very specialized in how they apply their medical skills in their perspective environments. As stated earlier, medics receive specialized military training, including basic soldiering skills and medical training. The scope of their medical training often surpasses the basic EMT level, incorporating trauma management, minor field surgery, and additional skills necessary for combat environments, like handling chemical injuries, performing under fire, and coordinating casualty evacuation. Combat medics are trained to manage traumatic injuries, such as gunshot wounds and blast injuries, often without the support of an entire medical team or facilities. On the other hand, EMTs are trained with a broad focus on civilian medical emergencies, including trauma, heart attacks, strokes, respiratory emergencies, and obstetrical emergencies, among others. Based on the National Association of Emergency Medical Technicians (NAEMT), their training includes patient assessment, emergency care, and transport procedures [6]. While they may encounter trauma injuries resembling those seen by combat medics, the frequency of such injuries is typically lower, and their training is less combat focused.

Finally, the scope of practice differs significantly between medics and EMTs due to their work environments. Often, medics work autonomously or with limited guidance in the field. They must rely heavily on their judgment to make critical care decisions under pressure. Their scope of practice is broader, allowing them to perform procedures like needle decompression or cricothyroidotomies in the field, which are interventions outside the usual purview of civilian prehospital care [7]. EMTs follow strict protocols and typically have direct lines of communication with physicians for guidance during their care. They operate within a more structured medical system, with readily available backup and transfer of care to advanced healthcare facilities. Their scope of practice, defined by local laws and regulations, often includes basic life support skills such as CPR, automated external defibrillator

(AED) use, oxygen administration, and medication delivery for specific emergencies [6].

While combat medics and EMTs share a common goal of providing emergency medical care, their roles, responsibilities, and experiences diverge significantly due to their different operational environments, the nature of their training, and their defined scope of practice. Appreciating these differences can contribute to a better understanding of the diverse roles these two healthcare specialties can provide to an HPO program.

Specific Applications and Metrics of the Medical Domain

As part of a human performance optimization program, physician assistants (PAs) and medics play a significant role in collecting and utilizing various tests and data to assess and enhance the performance and well-being of the HWS. This section focuses on the types of tests and data commonly collected by medical professionals in the medical domain, highlighting their importance in optimizing human performance:

1. Medical history and physical examinations: the medical team will gather comprehensive medical histories, including past illnesses, injuries, surgeries, and family medical history. This information helps identify potential risk factors and guide personalized care.
2. Laboratory and diagnostic tests: the team may order and interpret various laboratory tests, such as blood work (complete blood count, lipid profile, metabolic panel), urinalysis, and infectious disease screening. These tests help assess overall health, identify potential deficiencies or abnormalities, and guide treatment decisions. Furthermore, they may order diagnostic imaging tests, including X-rays, ultrasounds, and CT scans. Medical professionals use these scans to visualize internal structures, detect injuries, and diagnose conditions affecting bones, organs, or soft tissues.

Collecting and analyzing these tests and data allow medical domain personnel to develop comprehensive profiles of individuals, identify potential health issues, detect performance limitations, and tailor interventions to optimize human performance. This data-driven approach enables personalized care, the implementation of targeted interventions, and the continuous evaluation of progress toward optimal performance.

Conclusion

The medical domain is an essential component of the HPO program. PAs, medics, and EMTs undergo very stringent, specialized medical training. Their expertise brings access to healthcare closer to the HWS. Improving access to healthcare can help the HPO program become a preventative healthcare model, along with performance optimization.

References

1. Become a PA [Online]. [date unknown]. https://www.aapa.org/career-central/become-a-pa/ [13 July 2023].
2. https://online.shrs.Pitt.edu [Online]. [date unknown]. https://online.shrs.Pitt.edu [13 July 2023].
3. PANCE Requirements [Online]. [date unknown]. https://www.nccpa.net/become-certified/#pance-requirements. [13 July 2023].
4. Fedor PJ, Dorsch J, Kharod C, Paladino L, Rush SC. Rationale and implementation of a novel special operations medical officer course. J Spec Oper Med. 2021;21 https://doi.org/10.55460/y7jg-kp26.
5. Detro JF. Time for a change: recommended MTOE rank adjustment for Army Special Operations physician assistants. J Spec Oper Med. 2010;10:10–3. https://doi.org/10.55460/bpxw-kwiw.
6. NAEMT Education Standards [Online]. [date unknown]. https://www.naemt.org/education. [13 July 2023].
7. Medical Education and Training Campus. Combat medic specialist training program [Online]. [date unknown]. https://www.metc.mil/Academics/Programs/Combat-Medic-Specialist-Training-Program/. [13 July 2023].

Chapter 8
Spiritual Domain

Introduction

Spirituality refers to those beliefs, values, and practices that relate to the search for existential meaning, purpose, or transcendence, which may or may not include belief in a higher power [1]. When applying this concept to the overall readiness of a person or human weapon system (HWS), spirituality lacks tangibles. With this concept, it is hard to create tests to assess one's spirituality. To address this shortcoming, many organizations reframe the view of spirituality in the context of spiritual resiliency. HWS human performance optimization (HPO) programs typically measure spiritual resiliency through spiritual readiness or fitness.

Spiritual Readiness

Spiritual readiness, although a broad and often personalized concept, generally refers to an individual's preparation or capacity to leverage their spiritual or religious beliefs and practices to face, cope with, and overcome life's challenges and adversities. The terms spiritual readiness and spiritual fitness are often used interchangeably. For the rest of this chapter, only spiritual fitness will be used.

The Army field manual (FM) 7-22 defines spiritual readiness as "the ability to endure and overcome times of stress, hardship, and tragedy by making **meaning** of life experiences." The underlying foundation of spiritual readiness is one's intrinsic beliefs, values, motivations, and personal identity. Spirituality applies to religious and nonreligious personnel [2–4]. The downside to the FM 7-22 definition is that it is very individualized and lacks a worldwide view. This is the problematic side of this concept. Spiritual readiness is an individualized concept. Trying to apply this concept to an organization is difficult at best. Several US and NATO military

C. Myers, *The Human Weapon System*,
https://doi.org/10.1007/978-3-031-45061-7_8

organizations take a more inclusive view of spirituality readiness to address this obstacle [5–10].

The US Air Force defines spiritual readiness as "the proactive practice of establishing a sense of purpose or personal priorities to develop the skills needed in times of stress, hardship, and tragedy" [8]. On January 27, 2023, Captain Kevin Buss reported that US Navy Chaplain Lieutenant Commander Austin Grimes, chaplain of the 13th Marine Expeditionary Unit, stated: "Spiritual readiness is the ability of a warfighter to accomplish their mission with honor" [11]. Furthermore, Rear Admiral Terry Eddinger echoed Lieutenant Commander Austin Grimes' definition of spiritual readiness [5, 7]. However, Rear Admiral Terry Eddinger further wrote:

> It is the strength of spirit that helps a person get through adversity and stress with the ability to keep going. This is developed through a personal connection to a Higher Power and through the pursuit of **meaning, purpose**, values, connections to the sacred, and sacrificial service. In other words, warfighters are spiritually ready when they are connected to a source of **meaning** and **purpose** greater than themselves, able to explain why they serve, and are prepared to do their duty in peace and war. It is the epitome of honor, courage, and commitment. Spiritual readiness helps one perform and endure through a crisis [5].

Each of these definitions of spiritual readiness has a commonality. The essential qualities of spiritual readiness are one's concepts of **meaning** and **purpose**. In organizational settings like the military, emergency services, or healthcare, spiritual readiness allows individuals to lean on their spiritual beliefs and resources to maintain moral integrity, make ethical decisions, and sustain **resilience** in high-stress situations.

Spiritual readiness, cultivating a sense of purpose, faith, hope, and inner peace, is fundamental in building spiritual resilience. This resilience is akin to a buffer that helps individuals withstand and bounce back from life's inevitable adversities and traumas. By cultivating spiritual readiness, individuals develop a deeply anchored inner compass that guides them in life's turbulent seas. When confronted with crises, individuals with spiritual readiness lean into their belief systems, values, and practices for support, inspiration, and a sense of coherence, enabling them to reframe and constructively navigate their adversities. Thus, spiritual readiness promotes spiritual resiliency, contributing to overall well-being and enhancing one's ability to find meaning and strength, even amid life's most challenging circumstances.

Spirituality can help improve human performance in several ways as it can help individuals achieve a better sense of well-being, purpose, and balance in their lives [2, 12, 13]. For example, in 2023, the US Army reported a 37% reduction in suicides with holistic health and fitness (H2F) programs that strongly focus on mental and spiritual well-being [14]. While the specific effects can vary depending on the individual and their unique spiritual beliefs and practices, here are a few key ways in which spirituality might enhance performance:

1. Stress management: spirituality can provide a framework or set of practices for dealing with stress, which can significantly impact performance. Prayer, medita-

tion, and other spiritual practices can help induce a relaxation response, reducing stress hormones and promoting overall health [15].

2. Mental health: spirituality often involves meditation, mindfulness, and prayer, which can enhance mental health. These practices can help reduce anxiety and depression, promote positive feelings, and improve concentration and focus [16]. This can enhance cognitive function and overall performance in various domains, from school to work to athletics.
3. Sense of purpose: spiritual beliefs often provide a sense of meaning and purpose in life. This can serve as a source of motivation, pushing individuals to strive for their goals and perform at their best [16].
4. Resilience: spirituality can provide strength in adversity, enhancing resilience [3, 9]. Believing in a higher purpose or power can help individuals cope with difficulties and bounce back from failures, improving performance.
5. Community and support: many spiritual paths involve being a part of a larger community [9, 16, 17]. This sense of belonging can provide emotional support, which can help individuals perform better by reducing feelings of isolation and loneliness.
6. Emotional intelligence: spirituality often encourages introspection and self-reflection. This can lead to better emotional intelligence, which includes self-awareness, self-regulation, motivation, empathy, and social skills. In addition, higher emotional intelligence can contribute to better interpersonal relationships and improved performance in team settings.

The Spiritual Domain has many common traits with the Psychological and Cognitive Domains (Fig. 8.1). The Spiritual, Cognitive, and Psychological Domains will be treated as one domain if the workforce and funding are available. The Spiritual Domain can have the same amount of interaction with HWS no matter

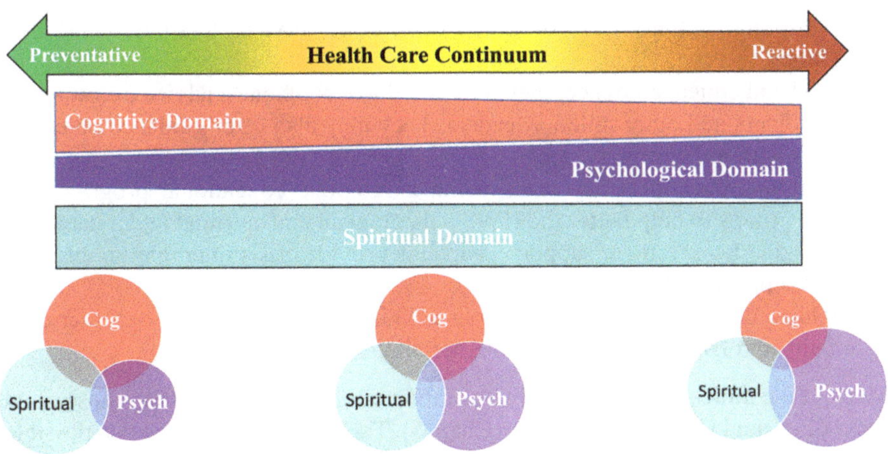

Fig. 8.1 The spiritual domain in the Healthcare Continuum. The spiritual domain will often work hand in hand with the psychological and cognitive domains

where the HWS is located on the Healthcare Continuum. The interaction and support are exact (Fig. 8.1). The Spiritual Domain support stays the same even though the level of interaction of the Cognitive and Psychological Domains change, as shown in the Venn diagrams in Fig. 8.1.

Spiritual Domain Team

This domain does not have the same structure from program to program. The military HWS professions tend to have a more standardized team structure. For example, the military Spiritual Domains, like those seen in the H2F, True North, and Preservation of the Force and Family (POTFF) programs, may have a chaplain or chaplain assistant.

Chaplain

Military chaplains serve in a unique way within armed forces around the world. They are commissioned officers who provide spiritual support, counseling, and moral guidance to service members, regardless of religious affiliation [18]. Here are some of their specific responsibilities and duties [18]:

1. Spiritual care: they conduct religious services, administer sacraments, and provide spiritual guidance to service members regardless of faith preference.
2. Pastoral care: they visit wounded soldiers in hospitals, support families of deceased service members, and may assist in notifying families of casualties or fatalities.
3. Counseling: military chaplains offer counseling services to soldiers and their families.
4. Moral and ethical guidance: they provide ethical and honest advice to commanding officers and other military leaders. They may also offer guidance on ethical issues within the military context.
5. Religious education: they often conduct religious educational programs and study groups to help better the HWS's theological and spiritual understanding.
6. Crisis and bereavement support: chaplains provide crisis intervention, bereavement support, and trauma counseling.
7. Building community: through various activities and programs, they foster community and friendship among HWS members.

Many of the duties and responsibilities of the military chaplain overlap those of the Psychology and Cognitive Domains (Chap. 5). These three domains, when possible, work hand in hand to bring the best possible care to the HWS. For example, HPO programs primarily focusing on social well-being and resilience will have all three domains, such as POTFF and True North.

Chaplain Assistant

The US Army and Air Force have occupational specialties named Religious Affairs Specialist and Religious Affairs AFSC, respectively. These positions are called chaplain assistants [19, 20]. The primary role of the chaplain assistant is to support the unit chaplain in providing religious services and support to HWS and their families. Additionally, they assist the chaplain in organizing religious services, ceremonies, and counseling sessions [19, 20].

As some of the responsibilities differ between the branches of service, some of the primary duties of a chaplain assistant are planning and conducting religious services, managing administrative tasks, assisting in counseling sessions, advising the unit commander on matters related to religious, ethical, and moral issues, and providing guidance on religious customs, traditions, and practices of different faiths [19, 20].

Nonmilitary Spiritual Domain Differences

Civilian HWS HPO programs do not have a religious support person assigned to their specific program. Instead, this type of support comes from the local community. However, it is up to the HPO program team to develop this relationship with the local church. The local church representative can provide the services described for the military chaplain; the military chaplain is an ordained minister! It is important to note that the nature and extent of collaboration between the local church and the HWS may vary depending on local circumstances, relationships, and the community's specific needs. Each church and HWS need to work together to determine the most effective ways to support the HWS's spiritual well-being while respecting the separation of church and state.

Spiritual Resiliency Tests

There is no universally accepted test to directly measure spiritual resiliency as it is an inherently subjective and deeply personal concept that can vary widely from person to person. However, a spiritual assessment typically assesses five different areas. These areas include "brevity, memorability, appropriateness, patient-centeredness, and credibility" [21–23]. However, several scales and inventories in positive psychology and spiritual studies can help assess constructs related to spiritual resiliency.

1. FACT [21, 22]
   (a) Chaplains developed the FACT scale. This test explores "F-Faith (and/or Belief); A-Active (and/or Available, Accessible, Applicable); C-Coping (and/or Comfort)/Conflict (and/or Concern); and T-Treatment" [21].

2. SWBS
   (a) The Spiritual Well-Being Scale test assesses two critical factors of spiritual resilience: one's sense of life purpose and life satisfaction [24].
3. SCS [25]
   (a) The Spiritual Coping Strategies scale is another tool that measures how individuals use their spirituality or religious beliefs to cope with stress, which can indicate spiritual resiliency [25].
4. MWoLQ [26–30]
   (a) The McGill Quality of Life Questionnaire was created to measure the quality of life for patients nearing the end of life caused by severe illness/disease [30]. In recent times, the MWoLQ has been adapted to measure spiritual resiliency.
5. IPOS [26, 31, 32]
   (a) The Palliative Outcome Scale is "a family of tools to measure a patient's physical symptoms, psychological, emotional and spiritual, and information and support needs" [31].
6. HOPE [24]
   (a) The HOPE concepts are as follows: "H" is for sources of "hope, strength, comfort, meaning, peace, love and connection"; "O" is for "the role of organized religion for the patient"; "P" is "personal spirituality and practices"; and "E" is for the "effects on medical care and end-of-life decisions" [33].
7. FACIT-Sp [34, 35]
   (b) The Functional Assessment of Chronic Illness Therapy-Spiritual Well-Being test covers 12 illness-specific items [35]. The FACIT-Sp was intended to "measure important aspects of spirituality, such as a sense of meaning and purpose in life and sense of the strength of comfort" [33]
8. TCI [36, 37]
   (c) The Self-Transcendence Character scale has three factors as part of the test. The factors include "self-forgetful vs. self-conscious, transpersonal identification, and spiritual acceptance versus materialism" [33]
9. BIO-PSSI [38, 39]
   (d) The biopsychosociospiritual inventory test uses a seven-symptom scale that concerns peace, harmony, and purpose [33]

10. Spiritual Orientation Inventory [40]

    (e) This test measures 85 items. These items are classified into nine subscales: "Transcendent Dimension, Meaning and Purpose, Mission in Life, Sacredness of Life, Material Values, Altruism, Idealism, Awareness of the Tragic, and Fruits of Spirituality" [33].

11. SIBS [41]

    (f) The Spiritual Involvement and Beliefs Scale contains 26 measured items. These items include "physicians conduct a quantitative spiritual assessment of patients and facilitate the study of patient spirituality in medical care" [33].

12. CSI-MEMO [21, 23, 42]

    (g) This measure includes Comfort (provided by religious beliefs), Stress (provided by religious beliefs), spiritual beliefs that might Influence medical decisions, Members of a religious or spiritual community and whether it is supportive, and Other spiritual needs that someone should address [21, 23, 33].

13. RSES [43]

    (h) The Response to Stressful Experiences Scale measures 22 items. The item scale includes "a brief, reliable, and valid measure of individual differences in cognitive, emotional, and behavioral responses to life's most stressful events" [33]. This test intends to measure resilience analysis of the five protective factors of meaning, active coping, cognitive flexibility, spirituality, and self-efficacy [33].

The above list is incomplete. These tests are not definitive and are subject to individual interpretation. They are tools to guide understanding rather than provide absolute measurements.

Conclusion

The Spiritual Domain is a very personal domain for many HWS professionals. Therefore, the scope and practice of this domain are highly dependent on the needs of the organization and the individual HWS. Military and civilian HWS organizations have different resources but can receive the same level of care. The most significant difference is that civilian HWS HPO programs need to see partnerships that will provide local religious organizations to meet the individual spiritual needs of their HWSs.

References

1. Hayden D. Spirituality in end-of-life care: attending the person on their journey. Br J Community Nurs. 2011;16:546–51. https://doi.org/10.12968/bjcn.2011.16.11.546.
2. Williams J, Ramsey V. The need for law enforcement wellness interventions: a critical review. The Sport Journal. 2017;24:1–18.
3. Steinhorn DM, Din J, Johnson A. Healing, spirituality and integrative medicine. Ann Palliat Med. 2017;6:237–47. https://doi.org/10.21037/apm.2017.05.01.
4. Missouri National Guard. Spiritual readiness [Online]. [date unknown]. https://www.moguard.ngb.mil/Programs/Holistic-Health-and-Fitness/Spiritual-Readiness/#. 28 July 2023.
5. Eddinger T. A frank discussion: readiness & resilience. Navy.mil; 2023.
6. Department of the Navy Strategic Plan for Religious Ministry [Online]. 2020. https://www.uscg.mil/Portals/0/seniorleadership/chaplain/DoN-Strategic-Plan-for-Religious-Mnistry-v-2-2020-02-07.pdf?ver=2020-02-21-105717-823. 29 July 2023.
7. Koenig H. Spiritual readiness in the military [Online]. https://www.youtube.com/watch?v=t7psbJYNigM&t=684s. 29 July 2023.
8. Air force handbook 36-2618 [Online]. www.e-Publishing.af.mil. 29 July 2023.
9. Koenig HG. "Spiritual readiness" in the U.S. military: a neglected component of warrior readiness. J Relig Health. 2023;62:1561–77. https://doi.org/10.1007/s10943-022-01563-z.
10. Spiritual Resiliency in the Canadian Armed Forces [Online]. http://www.journal.forces.gc.ca/Vol17/no2/PDF/CMJ172Ep26.pdf. 29 July 2023.
11. Buss K. How the Navy Chaplain corps safeguards marine readiness <United States Marine Corps Flagship>. News Display [Online]. 2023. https://www.marines.mil/News/News-Display/Article/3280196/how-the-navy-chaplain-corps-safeguards-marine-readiness/#. 29 July 2023.
12. Malmin M. Warrior culture, spirituality, and prayer. J Relig Health. 2013;52 https://doi.org/10.1007/s10943-013-9690-5.
13. Sax van der Weyden M, Toczko M, Fyock-Martin M, Martin J. Relationship between a maximum plank assessment and fitness, health behaviors, and moods in tactical athletes: an exploratory study. Int J Environ Res Public Health. 2022;19 https://doi.org/10.3390/ijerph191912832.
14. South T. Early data shows 37% suicide decrease in units with holistic health. Army Times. 2023. https://www.armytimes.com/news/your-army/2023/04/26/early-data-shows-37-suicide-decrease-in-units-withholistichealth/#:~:text=Data%20comparing%20soldiers%20in%20Holistic,suicides%20in%20non%2DH2F%20units.
15. Balducci L. Geriatric oncology, spirituality, and palliative care. J Pain Symptom Manag. 2019;57:171–5. https://doi.org/10.1016/j.jpainsymman.2018.05.009.
16. Weber SR, Pargament KI. The role of religion and spirituality in mental health. Curr Opin Psychiatry. 2014;27:358–63. https://doi.org/10.1097/YCO.0000000000000080.
17. Skoko I, Topić Stipić D, Tustonja M, Stanić D. Mental health and spirituality. [Online]. Psychiatry Danub 33: 822–826, 2021. http://www.ncbi.nlm.nih.gov/pubmed/35026808. 15 May 2023.
18. Carey LB, Hodgson TJ, Krikheli L, Soh RY, Armour AR, Singh TK, Impiombato CG. Moral injury, spiritual care and the role of chaplains: an exploratory scoping review of literature and resources. J Relig Health. 2016;55 https://doi.org/10.1007/s10943-016-0231-x.
19. Religious affairs specialist | goarmy.com [Online]. [date unknown]. https://www.goarmy.com/careers-and-jobs/career-match/support-logistics/admin-financial-support/56m-religious-affairs-specialist.html. 23 May 2023.
20. 5R0X1 – religious affairs AFSC – forever Wingman [Online]. [date unknown]. https://foreverwingman.com/career_fields/5r0x1-religious-affairs/. 23 May 2023.
21. Larocca-Pitts MA. FACT: taking a spiritual history in a clinical setting. J Health Care Chaplain. 2008;15:1–12. https://doi.org/10.1080/08854720802698350.
22. Thienhaus OJ. Spirituality in patient care: why, how, when, and what? Ann Pharmacother. 2002;36 https://doi.org/10.1345/aph.1c359.

23. Koenig H. Spirituality in patient care: why, how, when, and what. 1st ed. Philadelphia/London: Templeton Foundation Press; 2007.
24. Anandarajah G, Hight E. Spirituality and medical practice: using the HOPE questions as a practical tool for spiritual assessment. Am Fam Physician. 2001;1:81–9.
25. Hawthorne D, Youngblut JAM, Brooten D. Psychometric evaluation of the Spanish and English versions of the spiritual coping strategies scale. J Nurs Meas. 2011;19 https://doi.org/10.1891/1061-3749.19.1.46.
26. Selman L, Cert PG, Care P, Siegert R, Harding R, Gysels M, Speck P, Higginson IJ. A psychometric evaluation of measures of spirituality validated in culturally diverse palliative care populations. J Pain Symptom Manag. 2011;42 https://doi.org/10.1016/j.jpainsymman.2011.01.015.
27. Sguazzin C, Giorgi I, Alesii A, Fini M. Italian validation of the McGill Quality of Life Questionnaire (MQOL-It). G Ital Med Lav Ergon. 2010;32:B58–62.
28. Cohen SR, Sawatzky R, Russell LB, Shahidi J, Heyland DK, Gadermann AM. Measuring the quality of life of people at the end of life: the McGill Quality of Life Questionnaire-Revised. Palliat Med. 2017;31 https://doi.org/10.1177/0269216316659603.
29. Cui J, Fang F, Shen F, Song L, Zhou L, Ma X, Zhao J. Quality of life in patients with advanced cancer at the end of life as measured by the Mcgill Quality of Life Questionnaire: a survey in China. J Pain Symptom Manag. 2014;48 https://doi.org/10.1016/j.jpainsymman.2014.02.016.
30. Henry M, Huang LN, Ferland MK, Mitchell J, Cohen SR. Continued study of the psychometric properties of the McGill Quality of Life Questionnaire. Palliat Med. 2008;22 https://doi.org/10.1177/0269216308094519.
31. Cicely Saunders Institute. Palliative Care Outcome Scale. 2012.
32. Antunes B, Ferreira PL. Validation and cultural adaptation of the Integrated Palliative care Outcome Scale (IPOS) for the Portuguese population. BMC Palliat Care. 2020;19:178. https://doi.org/10.1186/s12904-020-00685-z.
33. Yeung D, Martin MT. Spiritual fitness and resilience: a review of relevant constructs, measures, and links to well-being [Online]. www.rand.org
34. Brady MJ, Peterman AH, Fitchett G, Mo M, Cella D. A case for including spirituality in quality of life measurement in oncology. Psychooncology. 1999;8:417–28. https://doi.org/10.1002/(SICI)1099-1611(199909/10)8:5<417::AID-PON398>3.0.CO;2-4.
35. Bredle JM, Salsman JM, Debb SM, Arnold BJ, Cella D. Spiritual well-being as a component of health-related quality of life: the Functional Assessment of Chronic Illness Therapy-Spiritual Well-Being Scale (FACIT-Sp). Religions (Basel). 2011;2 https://doi.org/10.3390/rel2010077.
36. Comings DE, Gonzales N, Saucier G, Johnson JP, MacMurray JP. The DRD4 gene and the spiritual transcendence scale of the character temperament index. Psychiatr Genet. 2000;10:185–9. https://doi.org/10.1097/00041444-200010040-00006.
37. Cloninger CR, Zohar AH. Personality and the perception of health and happiness. J Affect Disord. 2011;128:24–32. https://doi.org/10.1016/j.jad.2010.06.012.
38. Katerndahl DA. Impact of spiritual symptoms and their interactions on health services and life satisfaction. Ann Fam Med. 2008;6:412–20. https://doi.org/10.1370/afm.886.
39. Katerndahl D, Oyiriaru D. Assessing the biopsychosociospiritual model in primary care: development of the biopsychosociospiritual inventory (BioPSSI). Int J Psychiatry Med. 2007;37:393–414. https://doi.org/10.2190/PM.37.4.d.
40. Elkins DN, Hedstrom LJ, Hughes LL, Leaf JA, Saunders C. Toward a humanistic-phenomenological spirituality. J Humanist Psychol. 1988;28:5–18. https://doi.org/10.1177/0022167888284002.
41. Holland JC, Kash KM, Passik S, Gronert MK, Sison A, Lederberg M, Russak SM, Baider L, Fox B. A brief spiritual beliefs inventory for use in quality of life research in life-threatening illness. Psychooncology. 1998;7:460–9. https://doi.org/10.1002/(SICI)1099-1611(199811/12)7:6<460::AID-PON328>3.0.CO;2-R.
42. Koenig HG. An 83-year-old woman with chronic illness and strong religious beliefs. JAMA. 2002;288:487–93. https://doi.org/10.1001/jama.288.4.487.
43. Johnson DC, Polusny MA, Erbes CR, King D, King L, Litz BT, Schnurr PP, Friedman M, Pietrzak RH, Southwick SM. Development and initial validation of the Response to Stressful Experiences Scale. Mil Med. 2011;176:161–9. https://doi.org/10.7205/milmed-d-10-00258.

Part III
How to Build an HPO Program

Chapter 9
The Needs Analysis – Understanding the Needs of the Mission and the Human Weapon System

Introduction

Previous sections and chapters have referred to the needs analysis. This chapter will define and explain the process and provide an example of applying the needs analysis to an HWS HPO program. As you read this chapter, refer to Appendix A on the model needs analysis of an Arctic Infantryman from an MSK Domain point of view.

Needs Analysis – What Is It?

A needs analysis is necessary to determine the fundamental factors and training for a human weapon system (HWS) to succeed within the career field and the mission. This process identifies an HWS's career field, organization, and individual-specific strengths and limiters. As part of the process, the person performing the needs analysis will complete an individual, career field, and organization-specific injury analysis. This part of the needs analysis helps tailor the human performance optimization (HPO) program toward preventing such injuries. Overall, the primary aim of a needs analysis is twofold [1]: control and reduce injury and enhance HWS performance [1].

A proper needs analysis process involves three steps. The first step includes conducting the career/mission-oriented needs analysis. The second step comprises the HWS-oriented needs analysis. The third and penultimate step compares the career/mission and HWS needs analyses. The process looks as follows:

Step #1: Mission and organization-oriented needs analysis

(a) Needs of the mission and organization
(b) Injury analysis of the mission and organization

Step #2: HWS-oriented needs analysis/HWS profile

C. Myers, *The Human Weapon System*,
https://doi.org/10.1007/978-3-031-45061-7_9

Step #3: Comparative analysis of the mission and organization-oriented and HWS-oriented needs analysis

The following section will cover these three steps and use a real-world case study for the needs analysis.

Step 1: The Organization-Oriented Needs Analysis

An organization-oriented needs analysis helps the HPO team to fully understand the needs and demands of the HWS mission and organization. Defining the focus and direction of the HPO program is critical to the program's success. An appropriate organization-oriented needs analysis consists of two sections: (a) mission and organization characteristics and (b) injury analysis specific to that mission and organization.

This first part of the needs analysis assesses the various aspects and skills an organization needs to succeed in its mission. The elements considered in this analysis are as follows:

1. What is the Commander's mission, vision, and intent/key priorities?

   ****Note: This is the single most crucial step. If the HPO team does not do this, the needs analysis will be incomplete and untested to meet the organization's and HWS' needs.**
2. What are the organization's essential physical, psychological, and cognitive attributes?
3. How long is the average mission/training? How many missions/training exercises are conducted within a year? How much downtime exists between missions/exercises?
4. What tactics, techniques, and procedures (TTPs) are used?
5. What are the different sets of skills necessary to be successful in the mission? Are there general skills versus specialty skills?
6. What is the average experience level of the HWS in the organization? Are there enough experienced personnel in the HWS organization to properly train and mentor the junior personnel?
7. Do the training/missions occur on land, water, air, or a combination?
8. What are the clothing and equipment requirements? Does the equipment have any unique maintenance needs?
9. Are there any special medical requirements? What is the standard access level to care?
10. How many personnel are in the organization? How many different HWS professions are in the organization?

    (a) Many HWS organizations are comprised of HWS and non-HWS professionals. Identifying these different professionals helps understand the organization's needs. Furthermore, answering this question will assist in guiding Step 2 of the needs analysis.

11. What facilities and equipment are available? How much interaction time does the HPO team have with the organization's personnel?
12. What HPO domains are necessary to meet the organization's needs?
    (a) Section 2 only discussed the most common domains seen in HPO programs. Other domains can exist. Other possible consideration domains include life cycle management, education, and research. The bottom line is that the HPO team and the organization's leadership must collectively decide the domains necessary to meet the organization's needs.
13. What is the organization's annual budget for the HPO program? How many positions is the organization able to have?

The above list is not exhaustive. As each HPO Domain interprets the Commander's mission, vision, and intent, more questions can be generated. Each Domain will analyze these three concepts through their perspective professional lens. When the HPO team comes together to answer the organization's needs analysis question, it completes this section of Step #1.

The Organization-Oriented Needs Analysis: The Injury Analysis

The injury analysis section of the organization-oriented needs analysis looks at the common injuries involved with the mission and organization [1]. This part of the analysis does not consider injuries common to the HWS himself; the HWS-specific injury analysis occurs during the building and analysis of the HWS profile in Step #2. Some frequent questions investigated in this section of the process are as follows:

1. What are the current injuries associated with the mission and organization?
2. What is the injury frequency rate?
3. What is the average time loss with these injuries? What is the expected recovery time from the injury?
4. What are the common treatments for the injuries? What is needed to conduct the treatment?
5. Is there a particular time in the year when specific injuries occur?
6. Are there specific chronic injuries associated with the mission and organization?
7. Are there common predisposing factors for specific injuries?

Again, the above list is incomplete. The HPO team must come together to perform this analysis.

The organization-oriented needs analysis shows the demands to succeed and vital insight into common injuries. With this portion of the analysis, the HPO team can better conceptualize the mission and the organization's needs. This information helps lay the foundation for a complete and robust HPO program. However, first, the HWS's profile must be analyzed.

Step 2: The HWS's Needs Analysis/HWS Profile

This step depicts the goals, needs, and prevalent injuries associated with the HWS career field. This section in the needs analysis process provides specificity to the HPO program. As stated in Step #1, an HWS organization may have several different HWS professions. This aspect means an HPO team may have to perform Step 2 for each individual HWS profession within the organization. No matter how many different HWS professions exist, the process laid out in this section is the same. Furthermore, the entire HPO team creates the HWS profile together. The profile will be specific to each HWS profession. Performing this step for each HWS profession in the organization helps the HPO team better understand the individual needs of each separate HWS profession.

The HWS profile should, at a minimum, address the following areas listed below:

1. Goal setting
2. Injury analysis (physical, psychological, cognitive, emotional, spiritual, and other areas concerning the HWS)
3. Identification of the HWS's strengths and limitations across all domains
4. General and specific tasks
5. Kinematics and movement patterns specific to the career field
6. Experience level (i.e., early, mid, or late in career)
7. HWS body composition/body type
8. Medical history
9. Training availability
10. Psychological, cognitive, and spiritual needs

Goal setting is essential to the HWS profile; goals set the stage for what the HWS expects from the HPO program. Goals help guide, motivate, and provide buy-in for a program's participants. Setting goals can be a daunting and challenging process. One way to make it easier is to create attainable, realistic, and tangible goals. These three goal-setting criteria make goals achieveable. The last thing an HWS or HPO team would like to do is to create an unattainable goal. A challenging goal can be achieved with hard work. Setting the right goal(s) can set the stage for a strong HPO program and supply guidance on how the program should be structured.

The injury analysis is as critical as the goals. If nothing else is performed in this portion of the analysis, the injury analysis must be performed. Even though an injury analysis was conducted in Step #1, the mission and organization-specific injuries may differ from what is typically seen within the HWS career field.

Remember, organizations are typically comprised of multiple career fields. Each career field has specific needs. These needs are elucidated during this portion of the analysis. Understanding the specific injuries within a particular career field helps the HPO team tailor the HPO program to address those injuries. Refer to Appendix A for a detailed example of an HWS injury analysis.

Mental health is as vital as physical health in an HPO program. In high-stress environments, HWS professionals risk developing mental health issues like post-traumatic stress disorder (PTSD), anxiety, and depression. A needs analysis should involve mental health screenings and a candid assessment of work-induced stressors. These findings can shape the development of mental health support programs, including access to therapy, stress management training, and initiatives to promote a supportive work environment.

Aside from injury and PTSD, a needs analysis should also probe into other areas of concern, which may be specific to the organization or group. These might include nutritional deficiencies, sleep disorders, or cognitive performance issues. Each of these factors can significantly impact performance and should therefore be addressed in the HPO program.

An effective needs analysis is a powerful tool for uncovering the realities and complexities of an organization's human performance landscape. It reveals the strengths to be leveraged and the weaknesses to be mitigated. By addressing key areas such as injury, PTSD, and other specific concerns, the HPO program can be tailored to meet the group's unique needs, thereby ensuring maximum effectiveness and fostering an environment where every individual can perform at their optimal potential.

Identifying the HWS's strengths and limitations provides insight into skill proficiency. These skills are not limited to physical, technical, or tactical. The skills include cognitive, psychological, and nutritional skills. For example, how well does the HWS handle a crisis? Does the HWS know proper nutrition habits? Knowing the HWS strengths and limiters helps the HPO team understand where to focus their limited engagements. This information is gathered through performance testing, surveys, and interaction with each HWS. Obtaining this information forces the HPO team to interact with every HWS and critically analyze each data point. Knowing the HWS strengths and limiters does not mean the HPO programs focus solely on the HWS limiters. The program will work to make those limiters less of an obstacle for the HWS; however, the program must continue to improve on HWS strengths to ensure unit readiness and improved possibility for mission success.

The experience level of the HWS is a factor that must always be considered. The reasoning for this factor is twofold. The first reason is understanding the skill level and knowledge of the HWS. Here are examples:

1. Younger military personnel have experience with HPO teams, whereas older, more experienced military personnel may not. As a result, these members may need more training to understand new training and intervention paradigms and more time for buy-in.

2. Another way to view this first reason is through cognitive agility. The more experienced population tends to have a great capacity to move through "openness and focused attention" (aka cognitive agility) [2]. However, the younger population may not. Therefore, HWS population differences in cognitive agility is an area in which the Psychological, Cognitive, and Spiritual Domains can train to improve performance [3].

The second reason is injury rates. Again, this aspect ties into the injury analysis section of the HWS profile. Understanding how many and what injuries occur based on years of service/experience and age provides critical data to perform a cross-sectional injury analysis of the organization. This type of analysis can give the HPO team actionable data to address and prevent known injuries.

Anthropometric data provide insight into the HWS's body composition. These metrics are invaluable for understanding the individual HWS's body morphology and nutrition. Body composition is critical to understanding an individual's nutritional habits and level of physical fitness. Additionally, anthropometric information can be utilized as a data point to understand a possible underlying cause for specific injuries. The HPO program can use this information to understand the HWS better.

The next step is gathering the HWS's medical history. This history provides the HWS and HPO team with valuable information that might supply insights into HWS limitations and possible predisposing factors for injury. Furthermore, an excellent medical account provides previous diagnoses, vaccination status, medication history, allergies, etc. It allows the HPO team to understand the current health state of the HWS [4].

Individual and organization training availability (see Step #1) can differ. Many HWS organizations have other HWS professions. This aspect means there may be individual differences in working hours, workloads, etc. For example, an organization's training calendar may list performance training every day from 0700 to 0900 h. However, this does not mean every individual HWS will be in attendance. Some HWSs may be performing a mission or extended training, which precludes them from participating in the daily performance training. The MSK team should be aware of this aspect to plan accordingly. The same paradigm fits across all the HPO domains. Each domain team may have a tiny window for one-on-one or group sessions. This knowledge allows the HPO team to work with the organization and individual HWS to maximize limited interaction times.

The information gleaned from Steps #1 and #2 provides the framework for creating an organizational-level HPO program that can be tailored to the individual HWS needs. However, the needs analysis process is complete once these two steps are compared. This process occurs in Step #3, the comparative analysis.

Step 3: Comparative Analysis

Typically, the needs analysis is a two-staged process [1]. However, HPO programs require a third stage, the comparative analysis. A comparative analysis is when two or more variables are compared. The comparative analysis is where the HPO team puts all the puzzle pieces together to try to complete the HWS picture at the organizational and individual levels and allows the HPO team to compare the information gleaned from Steps #1 and #2 and create more profound insights. For example, taking the data from HWS experience levels (Step #2), organization (Step #1) and HWS (Step #2) injuries, and anthropometric data (Step #2) can be combined to create a cross-sectional and longitudinal analysis of injury rates. These two analyses can provide insights into when certain types of injuries occur (Fig. 9.2).

For example, Fig. 9.1 depicts ankle injuries within an HWS organization from 2019 to 2020. Unfortunately, this figure only provides a single dimension. Therefore, the only conclusion the HPO team can glean from Fig. 9.1 is that the HWS unit had 15 ankle injuries during the one year period. However, in Fig. 9.2, the comparative analysis provides more information. By conducting a cross-sectional analysis of the ankle injuries by rank and age, the HPO team can see that most ankle injuries in 2019–2020 occurred in enlisted members E1–E4 and ages < 30. Based on this more profound insight, the MSK team can target this segment of the HWS population with the proper interventions and education to try to prevent and reduce the number of ankle injuries.

The comparative analysis is part of the three-step needs analysis, where all the information is compiled. There is no set way to perform the analysis. The expertise and experience of each member of the HPO team drive the way the data are compared and analyzed. The contribution of each member of the HOP team provides a new perspective to the analysis and further completes the picture.

The final aspect is that the needs analysis process is not a magic bullet. It will answer only some of the questions. As a result, gaps in the data will be present. The

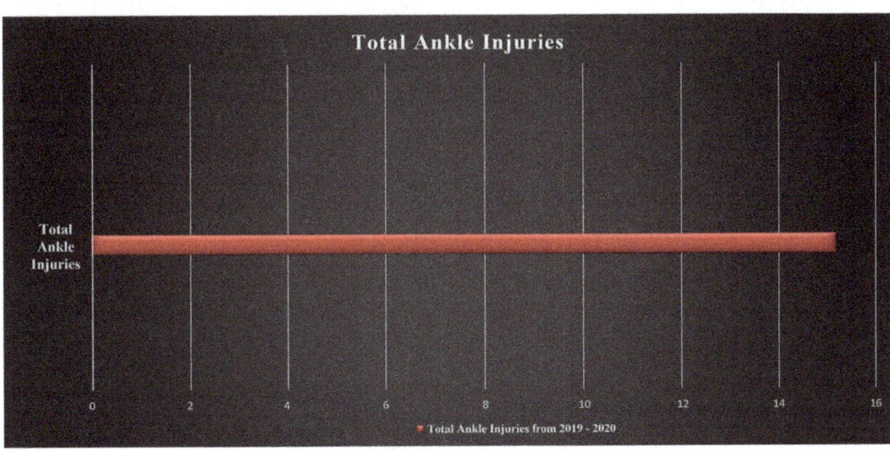

Fig. 9.1 Total ankle injuries for an HWS unit from 2019 to 2020

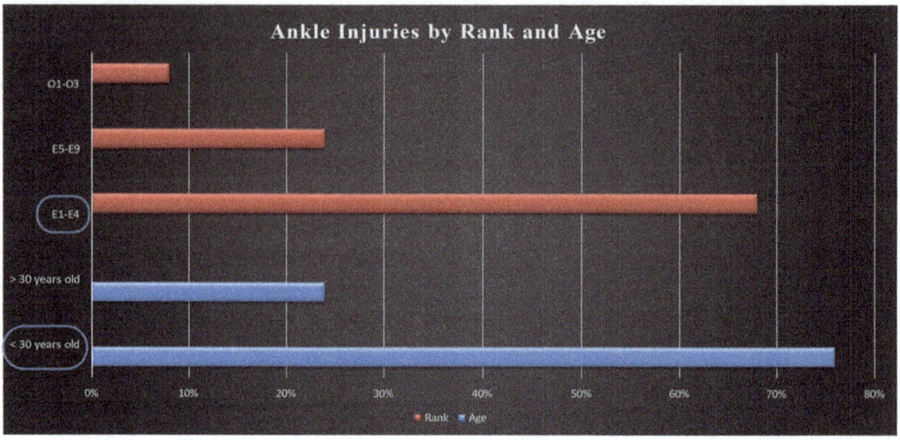

Fig. 9.2 Comparative analysis of ankle injuries of an HWS unit from 2019 to 2020

HPO team is responsible for identifying those gaps and trying to answer them as the program grows and develops.

To help you learn this process, Appendix A walks you through an example of how to perform a needs analysis. You are encouraged to look at this section and reread this chapter to gain a better understanding of this process. A proper needs analysis sets the tone and focus of a new HPO program and better sets the program for success in supporting the organization and HWS professions.

Conclusion

The needs analysis is a systematic approach that aids an HPO team in developing an individualized, comprehensive program. A proper HPO program will be able to address the strengths and limitations of every HWS within the organization. The methods of managing these items will be a mix of organizational and individual-level methods. The needs analysis process assists the HPO team to be able to identify these items and conduct a proper analysis.

References

1. Haff G, Triplett NT. Essentials of strength training and conditioning. 2014.
2. Ross J, Miller L, Deuster P. Cognitive agility as a factor in human performance optimization.
3. Ross J, Miller L, Deuster PA. Cognitive agility as a factor in human performance optimization. J Spec Oper Med. 2018;18.
4. Nichol JR, Sundjaja JH, Nelson G. Medical history [Online]. StatPearls Publishing. http://www.ncbi.nlm.nih.gov/pubmed/1148666. [26 Apr 2023].

Chapter 10
Periodization Methodology of the Human Weapon System

Christopher Myers and Kosta Telegadas

What Is Periodization?

Periodization is a training methodology used primarily in athletic and fitness programs to strategically plan and structure workouts to achieve peak performance at a specific time. It is based on dividing the overall training timeline (usually a year) into smaller, more manageable intervals, each focusing on a specific developmental goal.

The typical periodization model is divided into three primary cycles [1]:

1. Macrocycle: the most extended cycle, typically encompassing the entire training year or season, could be up to 4 years for an Olympic athlete.
2. Mesocycle: these are subsets of the macrocycle, typically lasting a few weeks to a few months. Each mesocycle has a specific focus, such as building endurance, improving strength, developing power, or peaking for a competition.
3. Microcycle: the minor cycle, usually a week, represents specific workouts and exercises performed within this short period to contribute to the goals of the mesocycle.

Within these cycles, periodization models often contain phases like the preparatory phase (general conditioning), competitive phase (specific training and competition), and transition phase (rest and recovery) [1].

C. Myers (✉) · K. Telegadas
3 Sports Science, Founder and Head Coach, Akron, OH, USA
e-mail: chris@3sportsscience.com

C. Myers, *The Human Weapon System*,
https://doi.org/10.1007/978-3-031-45061-7_10

Four main periodization models are used within human performance programs. These four models are linear, nonlinear/flexible, block, and conjugated. The following section details these periodization models [1].

Types of Periodization

Linear

Linear periodization is a training method used in body building to systematically manipulate training variables such as volume, intensity, and frequency over time to maximize performance gains [1]. The basic idea behind linear periodization is to divide a training cycle into distinct periods or phases, each focusing on different aspects of training. These periods typically range from a few weeks to several months and progress linearly, with each subsequent period building upon the gains made in the previous one [1].

The training program typically begins with a preparatory or foundational phase, which focuses on building a robust muscular endurance, flexibility, and technique base. This is followed by a hypertrophy phase, emphasizing high volume and moderate intensity training to increase muscle size and strength. The next phase is the strength phase, accentuating lower repetitions and higher intensity to build maximum strength. Finally, the program concludes with a peaking phase designed to optimize performance for a specific event or competition [1].

The overall goal of linear periodization is to optimize training adaptations by manipulating training variables progressively and systematically. By gradually increasing intensity and volume over time, the body can adapt to the demands of the training program, resulting in improved strength, power, and muscular endurance. An example of linear periodization is a follows:

Week 1: 4x5 Back Squat
Week 2: 4x6 Back Squat
Week 3: 5x6 Back Squat
Week 4: Deload/Recovery.

Block Periodization

Block periodization is a training method used in strength training and athletics that involves dividing the training program into distinct training blocks, each with a specific training focus [1–3]. Each block is designed to improve a particular quality or skill, and the blocks are arranged sequentially to create a training cycle. The premise behind block periodization is to train specific qualities or skills during a defined period while minimizing the interference of other training aspects. This approach can benefit HWSs who compete in sports with multiple seasons or have specific performance goals at different times [1–3].

A block training program typically begins with a preparatory block, which focuses on developing general fitness, addressing imbalances, and preparing the HWS for more specific training [1, 3, 4]. Several blocks of specialized training follow this, each focusing on a particular skill or quality, such as endurance, strength, power, or speed. Each block has a specific training emphasis, and the training variables, such as volume, intensity, and frequency, are manipulated to optimize adaptation to that particular focus. A training block can range from 6 to 12 weeks in length. The training focus and variables may change between blocks, allowing the HWS to progress in different aspects of their performance over the training cycle [1, 3, 4].

Block periodization aims to optimize training adaptations by focusing on specific qualities or skills during defined periods while minimizing interference from other training aspects. This approach can improve athletic performance and be particularly effective for HWSs with particular performance goals at different times of the year. An example of block periodization is as follows:

Block 1: GPP & Restoration
Block 2: Hypertrophy & Aerobic Conditioning
Block 3: Strength & Anaerobic Conditioning
Block 4: Repeat GPP & Restoration.

Nonlinear/Flexible Periodization

Flexible periodization is a training method that allows for greater customization and individualization of the training program based on an HWS's needs, goals, and response to training [1]. This periodization method is data driven [5]. Unlike linear and block periodization, which have a preset sequence of training phases, flexible periodization allows more flexibility in designing the training program [1, 6].

The training program is still divided into phases, like linear and block periodization. Still, the length and intensity of each phase may be adjusted based on the individual HWS's response to training and progress toward their goals [1]. The phases of flexible periodization may be organized around specific training goals, such as strength, power, or endurance, or they may be organized around competition schedules or other specific events. For example, if an HWS is not responding well to a particular training phase, the coach may adjust the length or intensity of that phase to suit the HWS's needs better.

Flexible periodization can be particularly useful for HWSs with unique training requirements or who may respond poorly to a more rigid training program. This is typically seen with the HWS. These professionals have variable schedules and limited training time that is friendly to strict periodized HPO programs. This approach allows HPO teams to tailor the training program to the specific needs of each HWS and adjust the program as needed. This can result in improved performance and reduced risk of injury as the program is better suited to the individual HWS's needs.

Conjugate Periodization

Conjugate periodization is a training method that combines the training of multiple physical qualities simultaneously throughout a training cycle [1]. It integrates the development of different qualities, such as strength, power, speed, and endurance, into the training program rather than focusing on a single quality during a specific training phase, as seen with linear and block periodization.

The critical principle of conjugate periodization is that each training session or week incorporates exercises or activities that target different physical qualities. This approach allows HWSs to progress in multiple areas simultaneously, leading to well-rounded physical development [1]. For example, the conjugate method enables a coach to work different variations of the core lifts of bench press, deadlift, and back squat. These core exercises are paired with accessory training.

In a conjugate periodization program, the training cycle is shorter than linear, block, and sometimes linear periodization. Conjugate training typically lasts 1–2 weeks. Within each microcycle, training sessions are designed to target different physical qualities. For example, one training session may focus on maximum strength, while another emphasizes power or speed. Each of these days will be coupled with different conditioning protocols as well. For example, the training protocols for the maximum strength day can focus on anaerobic conditioning with strength-based accessories. For the speed/power day, the focus might be on power endurance repeats towards the end of the training session with light load accessories for higher volume relative to the maximum strength day.

Conjugate periodization offers several benefits. It simultaneously allows HWSs to develop multiple physical qualities, promoting well-rounded athletic performance. By regularly varying the exercises and training methods, this methodology helps prevent adaptation plateaus and keeps the training program fresh and challenging. Additionally, it allows for individualization as exercises and training methods can be adjusted based on an HWS's specific needs and weaknesses and address injuries.

The four periodization training models help lay out how an HPO program can best improve and time HWS performance. Typically, the limiting factors of available training time and interaction with the HWS determine which periodization model is best to use. The following section investigates the causality of the time limitation.

Training Time in the Context of the HWS

The Annual Training Plan

Creating an annual training plan (i.e., the macrocycle) for performance involves setting goals, creating a detailed schedule of workouts, and regularly reviewing and adjusting the plan as necessary. HPO professionals and the HWS need to consider rest and recovery periods within the program. The following gives a brief overview of how an annual training plan is created:

1. Setting of goals: goal setting is arguably the most critical part of creating a training plan. The goals intrinsically motivate the HWS [5]. The goals must be specific, measurable, achievable, relevant, and time bound (SMART) [7].
2. Periodization: as the name suggests, it is a typical annual training plan that lasts for 1 year. However, it can last for up to 4 years, as seen with Olympians qualifying for the Olympics. Based on the periodization model used, the year is divided into specific periods or phases [1]. These can be broken down into macrocycles (annual plan), mesocycles (several weeks to a few months), and microcycles (weekly schedule) [1]. Each period should have a specific focus. The typical phases used are as follows:
   (a) Preparatory phase: this phase usually takes up the bulk of the plan and is where you build your general conditioning. Low-intensity but high-volume exercises are the focus here. The preparatory phase can further be divided into public and specific preparatory phases [1].
   (b) Competitive phase: during this phase, the intensity of the exercises increases and the volume decreases. This phase is geared toward preparing for a competition or achieving the primary goal(s). Techniques, speed, and other skills are often trained in athletic competitions [1].
   (c) Transition phase: this is the rest and recovery phase. This phase allows the body to heal and prevent burnout. Again, low-intensity and fun activities are recommended [1].
3. Progress monitoring: during the cycle, one should regularly check progress against one's goals. Training logs and performance tests are typically used to track progress [1, 5].
4. Revisions: as training progresses, adjustments need to be made. These could be due to changes in your schedule, health, or move toward your goals.

An annual training plan is an athletic approach typically applied to the HWS HPO program. As laid out in Chaps. 1 and 2, the HWS is not an athlete. This aspect is still valid in HPO programming for the HWS.

Why Traditional Periodization and Programming Do Not Work for the HWS?

As defined at the beginning of this chapter, traditional periodization is a strategy commonly used in fitness and sports training. It involves organizing training into blocks or phases, each with a specific goal. However, this strategy only sometimes works for the HWS. The training intensity and volume of periodization may not fit into the HWS schedule. This aspect is especially true for linear and block periodization. These two training paradigms are not optimal for HWS professionals for several reasons:

1. Variable and unpredictable demands: HWS personnel often face unpredictable physical demands. For example, military personnel may need to conduct a heavy ruck march one day, engage in close-quarters combat the next, then perform long-distance swims or high-altitude parachute jumps. Another example is firefighters. These HWS professionals conduct rescue operations as part of their primary job role. This includes extracting individuals from burning buildings, crashed vehicles, or collapsed structures. These tasks are physically and mentally exhausting in the military and firefighter professions. Traditional periodization assumes a predictable progression of training stimuli and does not account for the physical stimuli from the daily tasks of the HWS.
2. Constant readiness required: unlike athletes who prepare for specific events or seasons, HWS personnel must always be ready for various physical tasks. These professionals do not have "off-seasons" or periods where specific physical capacities might be de-emphasized in favor of others. The HWS is consistently in a competition phase. As described earlier, the competition phase is the phase of the highest training stress geared toward achieving athletic or physiological goals. However, any athlete, novice to elite, can only handle high training stress for short periods. It will break when the body is pushed too hard for too long. The HWS is always in a competition phase since the HWS cannot afford a down training period. Following a traditional periodization model, like linear and block, will cause too much training stress for the HWS. Many times, this is how injuries occur within HWS professions.
3. Physical vs. operational stress: traditional periodization focuses on managing physical stress from training. However, HWS personnel also face significant operational stress, including lack of sleep, irregular meals, high cognitive demands, and emotional stress. These factors can interfere with recovery and adaptation, making predicting how an individual will respond to a given training load harder. To help account for this aspect, utilize readiness scores and divide each training session into 25, 45, and 60-minute training blocks. Highlight the most important movements of the day for the 25-minute block, accessory movements in the 45-minute block, & and other movements that are not as important to said day in the 60-minute block.
4. Limited training resources: HWS personnel often need more time for training due to their operational duties. They may also need access to specialized equipment or facilities. Traditional periodization methods often assume that athletes have ample time to train and access various training resources. An HPO professional's programming within the HWS setting should create programs that can be scalable from bodyweight exercises, limited equipment (kettlebell & dumbbell only), and a fully equipped HWS facility similar to a Divison 1 Athletics facility.
5. Long-term health and function: Unlike athletes, who may be willing to accept increased injury risk or long-term health consequences in pursuit of short-term performance goals, HWS personnel must maintain their health and physical

function over a long career and into retirement. This aspect requires a balanced and sustainable approach to training than traditional periodization provides.

Given these factors, HWS professionals benefit from a more flexible and adaptable approach to training, such as nonlinear and conjugated periodization. The two periodization methodologies tend to have shorter block cycles, are data driven, and allow multiple capabilities to be trained simultaneously. Furthermore, these two methods are adaptive enough to allow operational stress to be considered part of the training cycle. Through the flexible and adaptive nature of nonlinear and conjugated periodization, the HPO program can better address the individual needs of each HWS, improve performance, and mitigate injury. However, for programming to work and periodization to be applied effectively, human beings need to come first. Data is only as reliable and valid as the frequency and consistency of the utilizations within the HPO process.

Conclusion

Periodization is a systematic training methodology for performance programs used to help athletes achieve their athletic and competition goals. Four methodologies exist: linear, nonlinear/flexible, block, and conjugate. Only nonlinear/flexible and conjugate periodization is optimal for the HWS. This aspect is due to variable and unpredictable demands, consistently required readiness, operational stress, limited training resources, and long-term health concerns. Flexible and conjugate periodization is data driven, a flexible training paradigm that can account for the daily stresses and unpredictability of the HWS to ensure improved performance and injury mitigation.

References

1. Bompa T, Buzzichelli C. Periodization: theory and methodology of training. 6th ed. Human Kinetics; 2018.
2. Stone MH, Hornsby WG, Haff GG, Fry AC, Suarez DG, Liu J, Gonzalez-Rave JM, Pierce KC. Periodization and block periodization in sports: emphasis on strength-power training – a provocative and challenging narrative. J Strength Cond Res. 2021;35.
3. Issurin V. New horizons for the methodology and physiology of training periodization: block periodization: new horizon or a false dawn? Sports Med. 2010;40.
4. Issurin VB. Benefits and limitations of block periodized training approaches to athletes' preparation: a review. Sports Med. 2016;46.
5. Myers C, Allen H. Triathlon training with power [Online]. 1st ed. Cognella Publishing. https://titles.cognella.com/triathlon-training-with-power-9781793532725. [11 Sept 2021].
6. McNamara JM, Stearne DJ. Effect of concurrent training, flexible nonlinear periodization, and maximal-effort cycling on strength and power. J Strength Cond Res. 2013;27 https://doi.org/10.1519/JSC.0b013e318274f343.
7. Boogaard K. How to write SMART goals (with examples). Work life by Atlassian: 2021.

Chapter 11
How to Determine If the HWS HPO Program Is Working

Introduction

Implementing a rigorous testing process in any training program cannot be overstated. Programmatic testing validates the program's efficacy and serves as a crucial feedback mechanism to continuously refine and improve the program. Testing uncovers vital information about the HWS's comprehension and skill acquisition and reveals gaps that need addressing for a more effective program. Furthermore, benchmarking provides quantifiable data to measure the program's success. Therefore, a comprehensive, well-planned testing strategy is critical to unlocking the full potential of any HPO program.

Program Testing

Testing a training program is of utmost importance in ensuring its success. It assesses the program's effectiveness in imparting the intended knowledge or skills. It is like a mirror that reflects the degree of the program working regarding its planned mission and focus. Moreover, testing offers valuable feedback that paves the way for improvement. Not all training programs are perfect from the outset, and testing helps identify the areas that need refining (Fig. 11.1). Through testing, the HPO team can determine what works well and what does not [1]. This process helps enhance the program's quality. The process is never ending. The needs of the HWS and the organization are constantly evolving. Therefore, the program must grow with its clients.

Program evaluations help uncover gaps. These gaps could be areas where the program needs to service the HWS correctly or might require additional resources.

C. Myers, *The Human Weapon System*,
https://doi.org/10.1007/978-3-031-45061-7_11

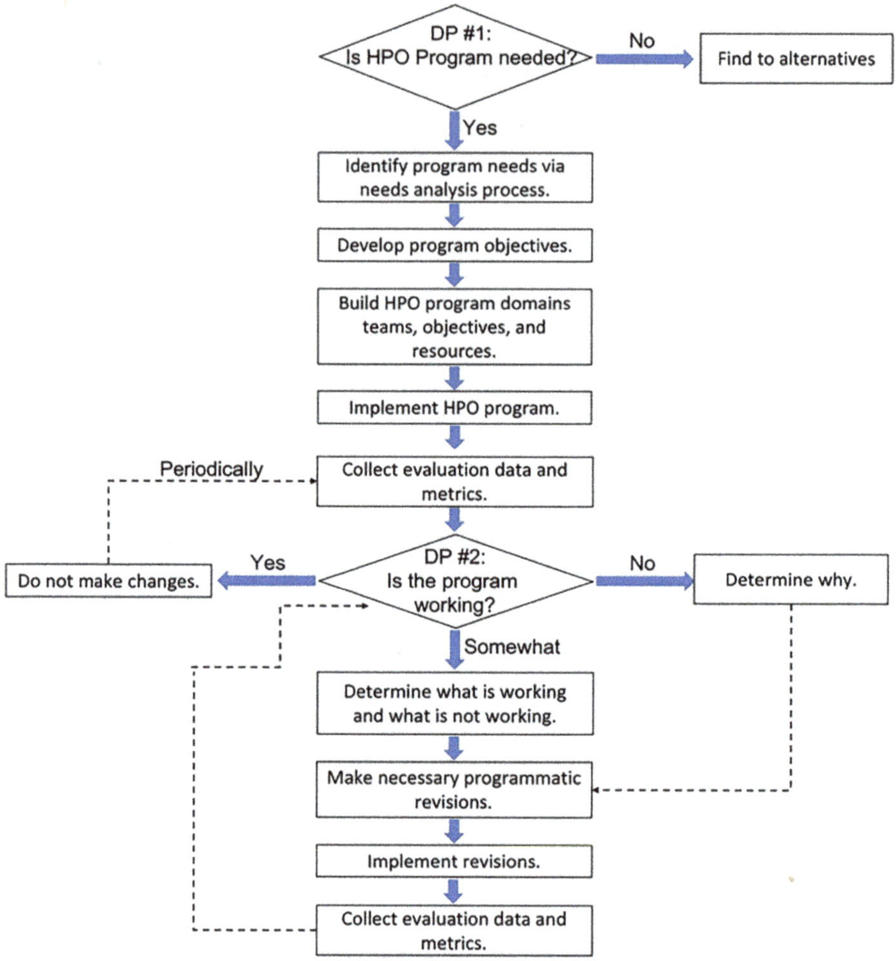

Fig. 11.1 Flowchart on how to test the viability of an HPO program. (This chart was adapted from Jenson et al. [1])

Identifying these gaps is critical to tailoring a more learner-centered program that fits the learners' needs and enhances their experience. The relevance of the content, effectiveness, and client satisfaction are essential aspects that should be evaluated when testing a training program. The HWS professions are dynamic, and what is relevant today might differ from tomorrow. Therefore, it is important to ensure that the training content stays current and relevant to the industry trends.

The Process: How to Get from Decision Point (DP) #1 to #2

Figure 11.1 details the process discussed throughout this chapter. The first step involves decision point (DP) #1, and that is for an organization to determine if an HPO program is needed. If the answer is yes, the program is established. Once the program is established, the HPO team must conduct a needs analysis (Chap. 9). The needs analysis helps the team to determine and build the HPO program domains, teams, focus, and resources. The program can be established after the program gathers the necessary resources, personnel, and funding. Once the program is functional, the HPO team needs to determine if the program is working. This aspect is conducted through data collection and analysis.

Data collection plays a pivotal role in measurement and reporting [1]. The HPO team must collect accurate and relevant data that illustrate the overall success and failures of the program. This is the only way to understand the program's efficacy truly. In Fig. 11.1, program metrics and data are collected and analyzed at several points in the testing process. During decision point (DP) #2, data and benchmarks provide the crucial information to answer the question, "Is the program working?" The HPO team should determine these metrics during the needs analysis (Chap. 9). The testing, analysis, and program revision is a feedback look process. It never ends. For certain HPO domains, testing is a requirement for compliance. This may be due to state law or certification agencies.

Benchmarks: What Are They and How to Use Them to Answer DP #2

Many performance programs utilize benchmarks to determine whether a program is working. Benchmarks are standards or points of reference against which things may be compared or assessed. In the realm of human performance, benchmarks serve as invaluable tools to gauge progress, identify areas of improvement that are relevant to the career field, and facilitate goal setting. Whether in athletics, academia, professional settings, or personal development, effective benchmarking can significantly enhance one's trajectory and outcomes.

Benchmarks are determined either through research or observations of a particular population. For example, the NSCA has created several strength and power benchmarks to gauge an athlete's ability against other athletes within that sport. Many of these benchmarks can be found in the **NSCA's Essentials of Strength and Conditioning, 4th ed**.

Figure 11.2 shows the benchmarks for 1RMs for the bench press, back squat, and power clean for the sports of basketball, softball, volleyball, and swimming. The benchmarks are based on the sample set of NCAA Division women participating in these sports. Furthermore, these lifts are known to correlate with how well an athlete performs within their sport. This type of data allows a coach or an athlete to see how she measures up to others in the sport. For example, swimming athlete A posts the following 1RMs for the aforementioned lifts:

TABLE 13.1 Percentile Values of the 1RM Bench Press, Squat, and Power Clean in NCAA Division I Female Collegiate Athletes

% rank	1RM bench press lb	1RM bench press kg	1RM squat lb	1RM squat kg	1RM power clean lb	1RM power clean kg	1RM bench press lb	1RM bench press kg	1RM squat lb	1RM squat kg
	BASKETBALL						**SWIMMING**			
90	124	56	178	81	130	59	116	53	145	66
80	119	54	160	73	124	56	109	50	135	61
70	115	52	147	67	117	53	106	48	129	59
60	112	51	135	61	112	51	101	46	120	55
50	106	48	129	59	110	50	97	44	116	53
40	102	46	115	52	103	47	94	43	112	51
30	96	44	112	51	96	44	93	42	104	47
20	88	40	101	46	88	40	88	40	101	46
10	82	37	81	37	77	35	78	35	97	44
Mean	105	48	130	59	106	48	98	45	118	54
SD	18	8	42	19	20	9	15	7	19	9
n	120		86		85		42		35	
	SOFTBALL						**VOLLEYBALL**			
90	117	53	184	84	122	55	113	51	185	84
80	108	49	170	77	115	52	108	49	171	78
70	104	47	148	67	106	48	104	47	165	75
60	99	45	139	63	100	45	100	45	153	70
50	95	43	126	57	94	43	98	45	143	65
40	90	41	120	55	93	42	96	44	136	62
30	85	39	112	51	88	40	90	41	126	57
20	80	36	94	43	80	36	85	39	112	51
10	69	31	76	35	71	32	79	36	98	45
Mean	94	43	130	59	97	44	97	44	144	65
SD	18	8	42	19	20	9	14	6	33	15
n	105		97		80		67		62	

lb = pounds, SD = standard deviation, *n* = sample size

Adapted, by permission, from Hoffman, 2006 (47).

Fig. 11.2 This figure depicts the benchmark values for NCAA Division 1 Women's Sports of basketball, softball, volleyball, and swimming [2]

1. Bench press = 108 lbs.
2. Back squat = 116 lbs.

Based on these results, athlete A placed in the top 70% for bench press and 50% for back squat for NCAA Division 1 Women Swimmers. This athlete needs to focus on lower leg strength and power, which could help improve her overall swimming performance.

As demonstrated in Fig. 11.2 and the above example, benchmarks provide an objective yardstick against measuring performance, eliminating ambiguity and ensuring clarity in evaluation. They provide clear targets that individuals can aspire to achieve, which helps with goal setting. Tracking progress against established benchmarks can be a significant source of motivation. Finally, regular benchmarking can highlight areas of weakness or where additional effort is required.

The difficulty for HPO programs is that benchmarks may not exist. The HPO team may need to create benchmarks relevant to their HWS population and program. In Fig. 11.2, the NSCA had to create these tangible competitive benchmarks by collecting the data from a sampling of NCAA Division 1 women athletes. For an HPO program, it may not be as straightforward. The HPO team must ask themselves, "What benchmarks are relevant to ensure the program works?" Each domain will have its own set of benchmarks. When each domain brings its benchmarks to the collective, the HPO program will have a way forward to answer DP #2 (Fig. 11.1).

Three main types of benchmarks exist. The first is industry standards. These benchmarks are common performance metrics that are widely accepted and recognized within professional and academic settings. The second is personal bests. These benchmarks the individual athlete's personal best performances. These types of benchmarks are extremely important for tracking personal development in sports. The third type is competitive benchmarking. This benchmark, as shown in Fig. 11.2, compares one's performance, typically personal bests, against others within that field or sport.

An HPO program and its domains can establish benchmarks through research. This is performed through understanding professional and academic standards, consultation, and population assessments. HPO professionals can research and interpret the professional and acceptable standards that exist within a particular domain. For example, Chap. 8 discusses Spirituality Resiliency tests. Each test has an acceptable metric that determines if an HWS is spiritually resilient or not. A similar metric exists for the MSK (i.e. Fig. 11.2), Nutrition, Medical, Cognitive and Psychological, and other HPO domains. The second option for creating a benchmark is to consult other experts within the perspective domain or field. Speaking with coaches, mentors, or industry leaders can gain insights into other valuable benchmarks that may not exist through research. The final option in creating benchmarks is through population assessments. Evaluating past performances of a particular HWS population can provide insights into what types of benchmarks are relevant and meaningful.

Making Benchmarks Meaningful Through Evaluation

Benchmarks are gathered by using technology (Chap. 12) and records; however, they are only helpful if they are meaningful. By making benchmarks meaningful, can they help the HPO program answer DP#2, "Is the program working?" Through consistent evaluation of the benchmarks, can the HPO team determine if the program is working and determine why or why not (Fig. 11.2). An accurate interpretation of results is essential. Many interpretations exist, but a simple, effective starting point is to look at the following three trends:

- Positive deviation: if one exceeds the standard benchmarks, this deviation is an excellent indicator of progress. However, the HPO team must ensure the positive change sustainable and not a one-off (i.e., a single personal best).
- Negative deviation: falling short of a benchmark can indicate areas needing improvement. These are the negative changes that need reflection on potential causes and the possibility of the need to adjust strategies.
- Consistency: while peaks and troughs are normal, consistent performance near the benchmark indicates stability and mastery.

What Comes After DP #2

The evaluation process is a never-ending evaluation loop. An HPO program should always try to improve itself. After evaluating the benchmarks, the HPO team should be able to determine if the program is working, somewhat working, or not. If the program works, the team should not change its current practices. However, if the program is somewhat or not working, the HPO team must determine what is not. Programmatic changes should be made and implemented. The implemented changes need to be evaluated after a certain period of time. As shown in Fig. 11.2, regardless of whether or not changes are implemented, the program is constantly evaluating itself through benchmarks to answer DP #2. Constant evaluation helps the program adapt to the ever-changing needs of the HWS.

Conclusion

Testing the HPO program is a vital component of the program's success. Collecting actionable and tangible data through benchmarking allows the HPO team to identify the program's strengths, weaknesses, and gaps. The testing process is never ending; an HPO program must continue to adapt and evolve. Only by critically analyzing the program can this action occur.

References

1. Jensen RC. Safety training: flowchart model facilitates development of effective courses [Online]. http://digitalcommons.mtech.edu/shih
2. Haff G, Triplett NT. Essentials of strength training and conditioning. 2014.

Chapter 12
Technology

Introduction

Various technologies (laboratory grade and wearable) and artificial intelligence (AI) exist to monitor and provide deep insights into human weight loss (HWS) performance. These technologies are necessary to maintain optimal physical, psychological, and cognitive performance at all echelons. In addition, information infrastructure and AI can help provide real-time assessments and decision-making. The current approach to measuring HWS performance is reactive; Health Professions Organization (HPO) teams need real-time data to make the best decisions possible.

HPO programs must leverage technology to measure physiological, nutritional, and cognitive aspects of human performance. This chapter will not discuss the efficacy of some of the available technologies. However, an HPO team member must ask, "Are we measuring the right thing?" This chapter introduces validated and used technology in successful HWS HPO programs, broken down into their respective domains.

Additionally, technology is used as human augmentation (HA). HA is "any material or non-material solution that can enhance an individual's ability to execute actions, tasks, and resilience in diverse/extreme environments" [1]. However, this chapter will focus not on technology for HA but on performance testing and monitoring purposes.

The HPO team and practitioners need to be aware of the claims technology companies can and will make about their products. Many times, the claims are grandiose and too good to be true. The team must research the technology and product to ensure it does what it claims to do. A popular claim is for a technology to predict injuries using machine learning and large data sets. This claim is unfounded and impossible to perform. The claims tend to come from small sample sets or limited movement patterns [2, 3]. If an HPO team utilizes technology, ensure the scientific literature supports the claims.

C. Myers, *The Human Weapon System*,
https://doi.org/10.1007/978-3-031-45061-7_12

MSK Domain

Technology in the MSK Domain mostly looks at performance progression and rehabilitation. Therefore, this section will focus on technologies used within the MSK Domain of an HPO program.

Velocity-Based Training/Accelerometers

Velocity-based training (VBT) is a weight-lifting method that uses real-time velocity feedback to adjust an exercise's load or volume [4, 5]. VBT is based on the principle that the velocity of a given lift can be used to estimate the intensity of the lift and the level of fatigue experienced by the lifter (Fig. 12.3).

In VBT, a lifter will use a specialized device (such as a velocity sensor or accelerometer) to measure the velocity of a given lift, such as a squat or bench press. These data then calculate the lifter's one-repetition maximum (1RM) and adjust the exercise's load or volume accordingly [4, 5].

For example, suppose a lifter aims to perform a set of squats at 85% of their 1RM but the velocity of their first few reps indicates that they are fatiguing quickly [4, 5]. In that case, the load may be decreased for subsequent sets to ensure that the lifter is still training at the appropriate intensity.

VBT has several potential benefits over traditional weight lifting methods, including improved accuracy in training load prescription, better monitoring of fatigue levels, and increased efficiency in training time. VBT has also improved athletes' strength and power output [4, 5]. However, it is essential to note that VBT requires specialized equipment and training and may only be suitable for some lifters. Additionally, like any weight lifting method, VBT risks injury if not performed correctly. Therefore, it is recommended that lifters consult with a qualified coach or trainer before incorporating VBT into their training program.

Fig. 12.1 The GymAware RS is an example of a tethered VBT sensor. (Photos courtesy of www.gymaware.com)

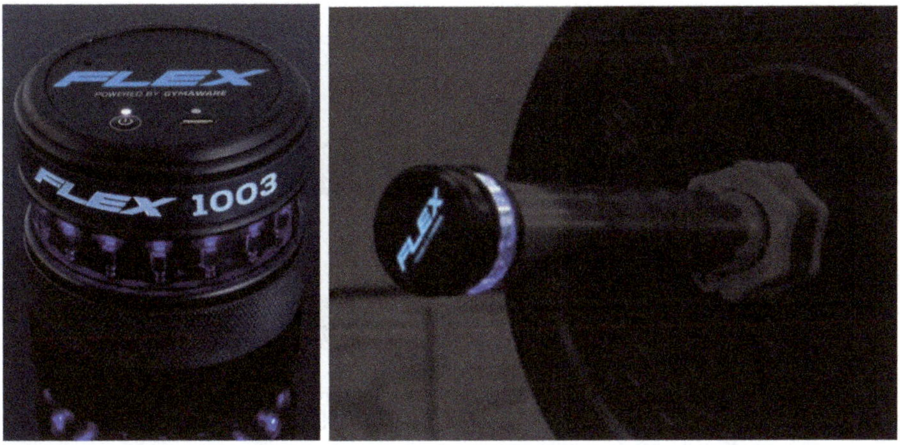

Fig. 12.2 The GymAware Flex is an example of a wireless VBT sensor. (Picture courtesy of https://www.flexstronger.com)

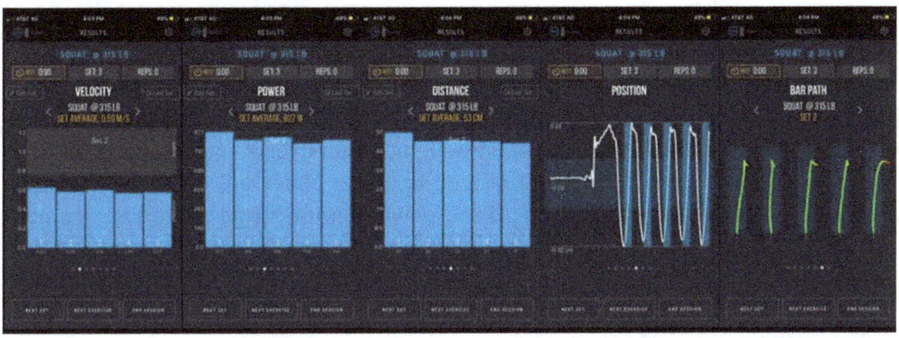

Fig. 12.3 This figure depicts the metrics available with VBT technology. This technology can provide metrics: velocity, power generated, total distance traveled, barbell position, and barbell path (left to right)

Currently, two types of VBT accelerometers exist: tethered and wireless. The tethered option (Fig. 12.1) has a cord directly attached to the barbell to measure speed. The wireless option (Fig. 12.2) uses a gyroscope and accelerometer technology to measure barbell speed (Fig. 12.3).

Force Plates

Force plates are "mechanical sensing systems designed to measure the ground reaction forces and moments involved in human movements" [6]. They are commonly used in biomechanics, sports science, and other fields where measuring the force

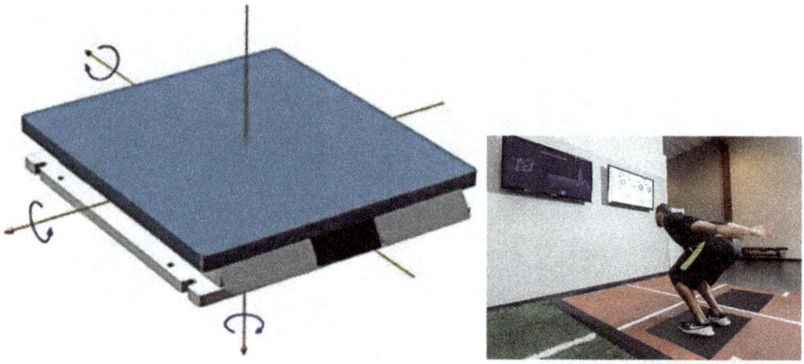

Fig. 12.4 (Left) This photo depicts the rotational forces a force plate measures [6]. (Right) The Sparta Science force plate is an example of force plate technology used to measure performance. (Photo courtesy of www.spartascience.com)

applied by an individual or object is essential. The basic design of a force plate consists of a flat surface fitted with sensors or transducers. When force is applied to the surface, the sensors measure the force and transmit the data to a computer or data acquisition system for analysis.

These devices can measure vertical, horizontal, and rotational forces (Fig. 12.4). They are often combined with motion capture systems to analyze human movements, such as gait analysis, countermovement jumps, or other sports-specific movements. For example, in HPO programs, force plates are commonly used to assess an athlete's lower limb strength and power output, range of motion, and neurocognitive ability and to track changes in performance over time through counter-jump movement and balance protocols. They can also be used to identify muscle imbalances and injury risk factors.

Force plates are vital for researchers and practitioners in many fields, especially in HPO. These devices provide valuable information about the forces involved in human movement and performance.

Metabolic Carts

A metabolic cart is a medical device used to measure a person's oxygen consumption ($\dot{V}O_2$) and carbon dioxide production ($\dot{V}CO_2$) during exercise or rest. This technique is termed indirect calorimetry. It evaluates a person's metabolic rate, which is the energy they burn at rest or during physical activity. This type of data provides essential insights into the HWS to develop personalized diet plans and training prescriptions. This device is commonly used in clinical and research settings to evaluate various conditions, including obesity, diabetes, heart disease, and lung disease. They are also used in sports science to evaluate HWS' endurance and fitness levels.

A metabolic cart typically consists of a mask placed over a person's nose and mouth to capture the air they breathe. The coach then analyzes the air to determine the person's oxygen and carbon dioxide levels. From these measurements, the cart can calculate the person's metabolic rate, which can be used to determine their energy expenditure, how many calories they are burning, and their fitness level.

Two types of metabolic carts are available, each with its own features and capabilities. Clinical and portable metabolic carts measure exercise metabolic performance and resting energy expenditure (REE). The type of cart used depends on the modality of exercise being measured and the available funds the HPO program has available. Some of the most common types include the following:

1. Portable metabolic carts: these carts are designed to be lightweight and easy to transport. They are typically used in field settings, such as sports science or military applications. The COSMED K5 is an example of this type of device (Fig. 12.5).
2. Clinical metabolic carts: these devices are used in hospitals and clinics to evaluate patients with metabolic disorders or diseases. These devices combine electrocardiograph (ECG) data with metabolic gas analysis (Fig. 12.6).

Overall, the type of metabolic cart used will depend on the specific application and the data that need to be collected.

Fig. 12.5 The COSMED K5. (Courtesy of www.cosmed.com)

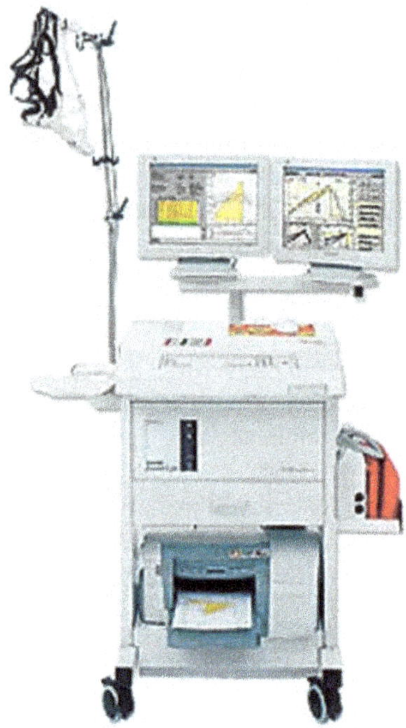

Fig. 12.6 An example of a medical-grade metabolic cart

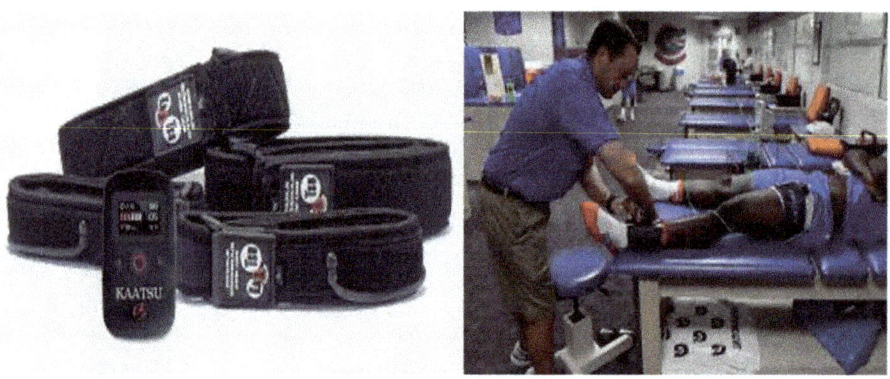

Fig. 12.7 (Left) This photo depicts a wireless BFR system produced by KAATSU. (Right) This photo depicts the use of a BFR in a PT recovery session

Blood Flow Restriction (BFR)

Blood flow restriction (BFR) devices are specialized tools used to restrict blood flow to a muscle during exercise partially (Fig. 12.7). BFR training aims to create a hypoxic (low-oxygen) environment in the muscle, which can lead to increased muscle growth and strength. These devices typically have a cuff or band wrapped around the limb (usually the upper arm or thigh). The cuff is inflated to a pressure high enough to partially restrict blood flow to the muscle but not wholly occlude it.

During exercise, the restricted blood flow causes a buildup of metabolic waste products in the muscle, such as lactate and hydrogen ions. This can increase muscle fiber recruitment and growth hormone release, promoting muscle hypertrophy (growth) and strength gains. BFR training is a well-validated training regime for improving muscle size, strength, and endurance. It is commonly used in rehabilitation settings, as well as in high-performance sports training.

For those who use BFR, it is essential that this training be performed under the supervision of a trained professional. BFR devices can cause injury or other complications. Furthermore, this training type is not suitable for everyone, particularly those with underlying health conditions or who are pregnant, due to blood flow restriction.

Gait Analysis

A gait analysis is a diagnostic procedure that assesses how a person walks, runs, or carries symmetrical and asymmetrical loads (Fig. 12.8). It is typically performed by a medical professional, such as an exercise physiologist, physical therapist, or orthopedist. It involves analyzing the mechanics of the person's gait to identify any abnormalities or issues affecting their mobility or causing pain.

During a gait analysis, the patient is typically asked to walk or run on a treadmill or walkway while filmed from multiple angles. The video footage is then analyzed to evaluate the person's gait, including the positioning and movement of their feet, legs, hips, and other joints. Other techniques may also be used during a gait

Fig. 12.8 Photo of a standard 16-camera gait analysis system. This is the type of system found in a gait analysis laboratory. (Photo Courtesy of BTS Bioengineering Corp., Brooklyn, NY, USA)

Fig. 12.9 This photo portrays the RunDNA gait analysis system. The RunDNA system is a portable three-cameral gait analysis system. (Photos courtesy of www.rundna.com)

analysis, such as pressure sensing mats or force plates, to measure the forces applied to the feet and legs during the gait cycle.

Gait analysis results can help identify various issues, such as muscle imbalances, joint problems, and poor biomechanics. Based on these findings, the medical professional may recommend specific interventions, such as physical therapy, orthotics, or changes to the patient's footwear or gait mechanics (Fig. 12.9).

Overall, gait analysis is a valuable tool for diagnosing and treating a wide range of mobility and movement-related issues and can help improve HWS performance and mitigate injuries.

Heart Rate (HR) Monitors and Smartwatches

Heart rate monitors are devices that track a person's heart rate, which is the number of times the heart beats per minute. Heart rate monitors are typically stand-alone devices worn on the wrist, chest, or other body parts. They provide real-time heart rate data during exercise or other activities and are often used by athletes or fitness enthusiasts to track their training intensity and recovery. These devices use different heart rate sensors, such as a heart rate monitor strap, a two-lead electrocardiograph (ECG), or near-infrared spectroscopy (NIRS).

Smartwatches, on the other hand, are multifunctional devices that typically include a heart rate monitor as one of their features. In addition to measuring heart rate, smartwatches can track various other data points, such as steps taken, calories burned, run pace, distance, and sleep quality. Smartwatches can also provide notifications, phone calls, and other features, making them versatile and convenient devices for many people.

Both heart rate monitors and smartwatches can be used to monitor heart rate during exercise, which can help track fitness goals and training stress, monitor recovery, and avoid overexertion. Additionally, some heart rate monitors and smartwatches have advanced features, such as GPS tracking or connecting with other devices, which can provide even more data and insights into the user's health and fitness.

Heart rate monitors and smartwatches are valuable tools for monitoring heart rate and tracking fitness progress. While they have slightly different features and use cases, both devices can provide useful information and help people achieve their fitness goals.

Nutrition

Nutrition technology focuses on using technological tools and solutions to help individuals improve their nutrition and overall health. Many nutritional technologies can be utilized in the MSK and Nutrition Domains. This aspect returns to the relationship between these two domains, as discussed in Chap. 5. This section will focus on technologies seen in the Nutrition Domain of an HPO program.

Body Density Measuring Technology

Body density technology is a method of measuring a person's body composition based on the principle that different tissues in the body have different densities. This method typically involves measuring a person's body weight and volume and then using these measurements to calculate their body density.

Several techniques are used in body density technology. The following are short descriptions of the available body density technology:

1. Dual-energy X-ray absorptiometry (DEXA or DXA): the DEXA system is considered the gold standard. DEXA is a technology used to measure bone density and body composition. It uses two low-dose X-ray beams of different energies to scan the body and produce bone and soft tissue images. A computer then analyzes the images to determine bone density and body composition (Fig. 12.10).

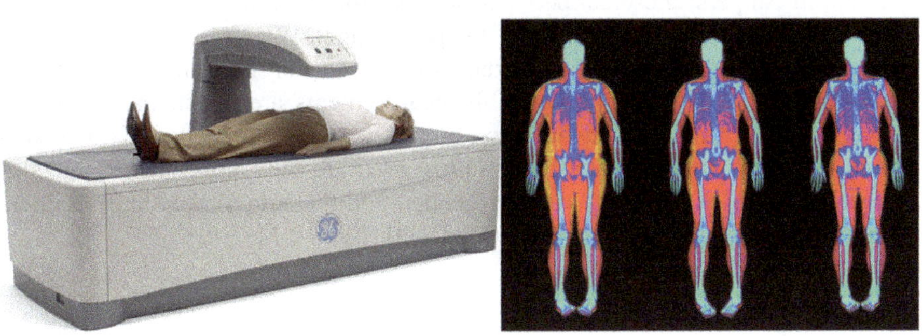

Fig. 12.10 The DEXA system. (Photo courtesy of www.gehealthcare.com)

2. Bioelectrical impedance analysis (BIA) technology: BIA is a method used to estimate body composition by measuring the resistance of body tissues to the flow of a small electrical current. It works on the principle that different tissues in the body conduct electricity differently. A small electrical current in BIA is sent through the body via electrodes placed on the skin. The current passes through the body's fluids, including blood and muscle, but encounters resistance when it passes through fat. By measuring resistance to the current, the BIA device can estimate the amount of body fat and lean muscle mass.
3. Hydrostatic weighing: this method, also known as underwater weighing, measures body composition based on Archimedes' buoyancy principle. This method involves the person being weighed while completely submerged in water. The body density calculation is used since fat has a lower density than muscle and bone. The more fat a person has, the less dense they are, and the more they will float. Lean muscle, bone, and other dense tissues sink more in water and have a higher density.

Once the person's body density is determined, it can be used to calculate their body fat percentage using a prediction equation. The most used equation is the Siri equation [7–9]:

$$\text{Body Fat Percentage} = \left[\left(4.95 / \text{Body Density}\right) - 4.5\right] \times 100$$

4. Air displacement technology: air displacement plethysmography (ADP) is a method used to measure body composition based on the principle of Boyle's law, which states that the pressure and volume of a gas are inversely proportional to each other [10–12].

During ADP testing, the measured person sits inside a small chamber called a Bod Pod. The Bod Pod measures the person's body volume by determining the amount of air displaced when the person is inside the chamber. The body density is calculated using the Siri equation (as with hydrostatic weighing). This method of body density measuring is more expensive than hydrostatic weighing but is considered more accurate.

5. Skinfold calipers: these calipers perform skinfold measurements, a commonly used method for estimating body fat percentage [13]. This method involves using a caliper to measure the thickness of the skin and subcutaneous fat at various locations on the body. Once all measurements are taken, the values are entered into a prediction equation to estimate body fat percentage. Many prediction equations are available, but the most used are the Durnin and Womersley equations, which are based on age, gender, and the sum of the skinfold thicknesses at various locations on the body [7–9] (Fig. 12.11).

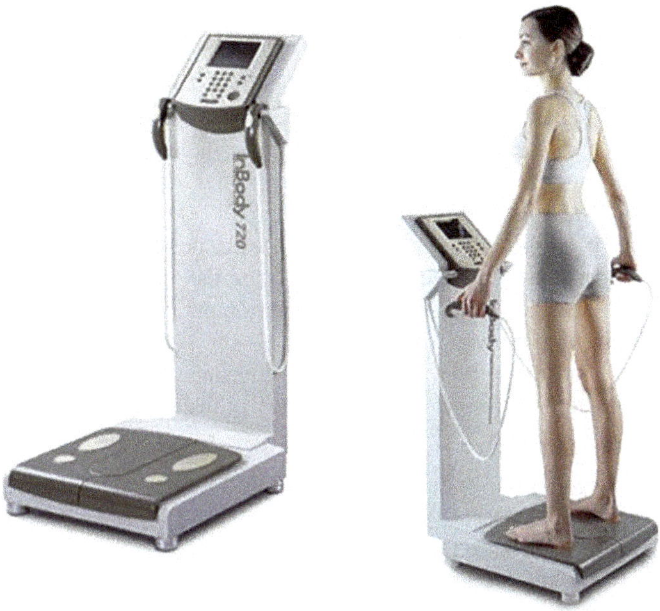

Fig. 12.11 The InBody system is an example of BIA technology

Nutrition Tracking

Nutrition tracking software is a computer program that allows individuals to monitor their daily intake of food and beverages and track the nutrients they consume. This type of software is typically designed to help people achieve peak performance, lose weight, build muscle, or maintain a healthy lifestyle. Nutrition tracking allows users to input the food they eat throughout the day and the portion sizes. The software then uses a nutritional information database to calculate each food item's calories, macronutrient content, and other vital nutrients, like vitamins and minerals.

Nutrition tracking software can be used on a computer or mobile device, and many apps integrate with other fitness and health tracking tools, like wearable fitness trackers or smartwatches. This software provides dietitians and the MSK team insights into how individual HWSs stay fueled and hydrated. Furthermore, dietitians can use this information to create meal plans to meet the needs of individual HWSs.

Some examples of nutritional tracking software are as follows:

1. Notemeal – www.notemeal.io
2. MyPlate – www.myplate.gov
3. Nutritionist Pro Diet Analysis Module – https://nutritionistpro.com/diet-nutrition-analysis

Biowearables

Biowearables are a type of wearable technology that is designed to monitor various biological functions and activities of the human body. These devices are typically worn on the body and can track multiple metrics, including heart rate, sleep patterns, blood pressure, blood glucose, lactate, hormones, etc. These sensors and other types of technology collect data about the wearer's body, which can then be analyzed and used to provide insights into their health, performance, and well-being. For example, individuals can use this data to track their progress toward fitness or health goals, or healthcare providers can monitor patients remotely and provide personalized treatment plans.

Biowearables include smartwatches, Oura Rings, and Whoop straps that monitor heart rate, sleep, and physical activity. Other biowearable devices include the following:

1. Continuous glucose monitors (CGMs) follow blood sugar levels in real time (Fig. 12.12).
2. Lactate monitors measure blood lactate levels.
3. Hydration and sweat monitors measure hydration levels.
4. Cortisol sensors measure the hormone cortisol (Fig. 12.13).

Cognitive

Cognitive training and performance technology refer to various computer-based programs and tools to enhance cognitive abilities, such as memory, attention, problem-solving, and decision-making. These tools can improve cognitive performance in multiple settings, including education, sports, and the workplace.

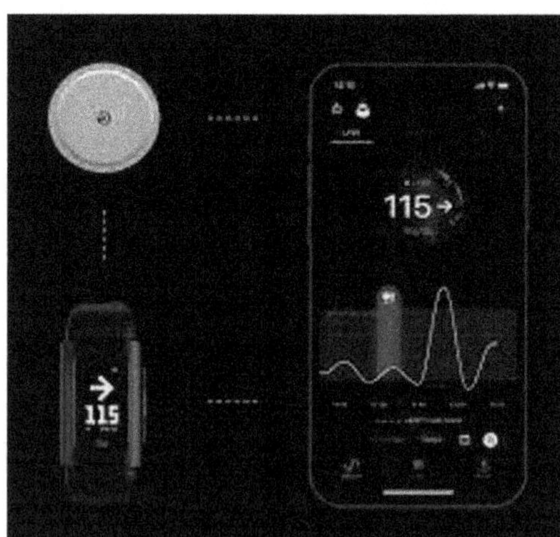

Fig. 12.12 This photo shows the Supersapien CGM system. (Photo courtesy of www.supersapiens.com)

Fig. 12.13 The Nix hydration sensor is an example of a commercial off-the-shelf hydration sensor. (Photo courtesy of https://nixbiosensors.com)

Fig. 12.14 Performing physical training in PPE increases the stress the HWS experiences. This type of training helps improve cognitive abilities under physiological stress

On the other hand, performance cognitive technology typically focuses on enhancing cognitive abilities in specific contexts, such as sports or the workplace. For example, cognitive performance technology may train athletes to improve their reaction times or decision-making skills or help workers in high-pressure environments make more effective decisions.

Cognitive training systems can be very simple to very complex. One way to improve cognitive ability is to perform a physical training workout while wearing personal protective equipment (Fig. 12.14).

Other elegant examples of cognitive training systems are as follows:

1. Simulators: a simulator is a computer program or system designed to imitate or replicate real-world scenarios or processes. The purpose of a simulator is to provide a safe and controlled environment for users to learn, practice, or test skills, procedures, or systems without the risks or costs associated with the real thing. This technology can be used in various contexts, including aviation, military, healthcare, engineering, and transportation. For example, pilots can use flight simulators to practice takeoff and landing procedures, and firefighters can use simulators to practice firefighting procedures.
2. Virtual reality (VR): VR is a computer-generated simulation of an environment or experience with which the user can interact seemingly naturally or physically. It typically involves using a headset or other device that tracks the user's movement and adjusts the virtual environment, creating a sense of immersion and presence.

VR technology involves hardware and software, including powerful computer processors, high-resolution displays, motion sensors, and specialized VR software. Together, these components create an immersive three-dimensional (3D) environment for the user to explore and interact.

3. Reaction training devices. Reaction training is a type of training that is designed to improve an individual's ability to react quickly and accurately to a stimulus. This type of training is often used in sports, military, and emergency response settings, where quick reactions can mean the difference between success and failure. This type of training typically involves exercises or drills that require the individual to respond quickly and accurately to a visual, auditory, or physical stimulus. For example, in sports training, reaction drills may involve a coach throwing a ball at a player, who must react quickly to catch it. In military training, reaction drills may include responding to simulated enemy fire or other threats. The benefits of reaction training include improved reflexes, better decision-making skills under pressure, and improved coordination and motor skills.

Reaction training can also involve using technology, such as lights, sensors, or virtual reality, to simulate real-world scenarios and provide feedback on reaction times and accuracy (Figs. 12.15 and 12.16).

Fig. 12.15 The Conflict Kinetics system is a synthetic gunfighter simulator that trains police and military forces in small-arms operations. (Photo courtesy of www.conflictkinetics.com)

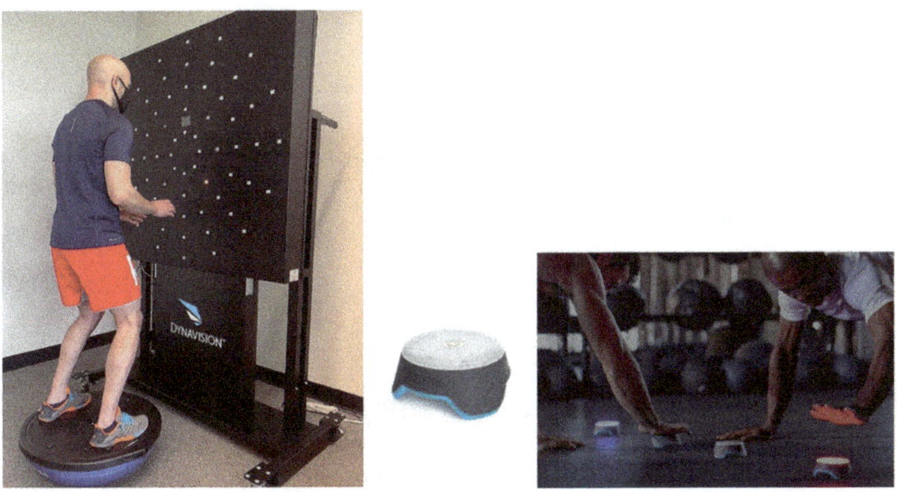

Fig. 12.16 The photos depict reaction drill technology utilizing light patterns. On the left is the Dynavision system [14], and on the right is the BlazePod system. (Photo courtesy of www.blazepod.com)

Psychology

In recent years, technology has helped the field of psychology to improve patient access and care. This section will focus on critical technologies seen within the Psychology Domain.

Telecommunication

Telecommunication applications such as Zoom and Teams have improved patient access to qualified professionals in the field of psychology. Traditionally, a patient must meet with the therapist at an office. Now, the patient can meet with a therapist in their own home. This aspect improves comfort and reduces stress for the patient.

Virtual Reality (VR)

VR has benefits in this domain as well as the Cognitive Domain. This technology can treat posttraumatic stress disorder (PTSD) and anxiety.

Biowearables

Biowearables, as described earlier, can provide a therapist with quantitative data on the well-being of a patient. This information can help the therapist conclude and determine if treatment is helping the patient.

Conclusion

Technology can provide objective data to the HPO team to improve the program and decisions that affect the HWS. This chapter should have discussed the effectiveness and quality of the data this type of technology provides. The focus of this chapter was to introduce the types of technologies currently being leveraged by successful HWS HPO programs. The HPO team must decide on what metrics to measure and the effectiveness of those metrics. The needs analysis process will help the HPO team determine which technologies best support the HPO program.

References

1. Haggenmiller C. Human performance optimization and enhancement. 2021.
2. Bullock GS, Mylott J, Hughes T, Nicholson KF, Riley RD, Collins GS. Just how confident can we be in predicting sports injuries? A systematic review of the methodological conduct and performance of existing musculoskeletal injury prediction models in sport. Sports Med. 2022;
3. Seow D, Graham I, Massey A. Prediction models for musculoskeletal injuries in professional sporting activities: a systematic review. Transl Sports Med. 2020;3
4. Signore N. Velocity-based training: how to apply science, technology, and data to maximize performance [Online]. https://www.amazon.com/Velocity-Based-Training-Technology-Maximize-Performance/dp/1492599956/ref=sr_1_2?keywords=velocity+based+training&qid=1683483194&s=books&sprefix=velocity+based+%2Cstripbooks-intl-ship%2C156&sr=1-2. [7 May 2023].
5. Tumminello N. Strength zone training: the most effective method for maximizing muscle development [Online]. https://www.amazon.com/Strength-Zone-Training-Nick-Tumminello/dp/1718211473. [7 May 2023].
6. Popovic MB, Lamkin-Kennard KA, Popovic MB. Sensors: natural and synthetic sensors.
7. Thompson WR, Gordon NF, Pescatello LS. ACSM's guidelines for exercise testing and prescription [Online]. 8th ed. Lippincot, Williams, and Wilkins. https://www.lww.com/Product/9781609139551.
8. Ehrman JK. ACSM Guidelines for Exercise Testing and Prescription. 8th ed; 2010.
9. Manning JM. ACSM resource manual for guidelines for exercise testing and prescription. J Athl Train. 1999;34:73–4. https://doi.org/10.1136/bjsm.25.4.244.
10. Powers SK, Howley ET. Exercise physiology: theory and application to fitness and performance. 8th ed. New York City: McGraw-Hill Humanities/Social Sciences/Languages; 2011.
11. Costanzo LS. Physiology [Online]. 7th ed. https://www.vitalsource.com/products/costanzo-physiology-linda-costanzo-v9780323793346. [20 Mar 2022].
12. Tortora GJ, Derrickson B. Principles of anatomy and physiology. [date unknown].
13. Jung MH, Namkoong K, Lee Y, Koh YJ, Eom K, Jang H, Jung W, Bae J, Park J. Wrist-wearable bioelectrical impedance analyzer with miniature electrodes for daily obesity management. Sci Rep. 2021;11 https://doi.org/10.1038/s41598-020-79667-3.
14. Rentz LE, Brandmeir CL, Rawls BG, Galster SM. Reactive task performance under varying loads in division I collegiate soccer athletes. Front Sports Act Living. 2021;3:278. https://doi.org/10.3389/fspor.2021.707910.

Arctic/Extreme Cold Weather Operations Needs: Analysis of the 7th Arctic Infantry Company

****Note: The information from this needs analysis was taken from open-source data from various Arctic units. The named infantry company used throughout this analysis is fictional.**

Introduction

Knowing, understanding, and applying personal metrics to training is only part of the training process. Using these metrics is critical to creating an effective training program. However, how does one get to this point? The first place to start is a systematic "needs analysis" approach. This systematic process can create a personalized and individualized program that trains the correct metabolic pathways and addresses the Human Weapon System (HWS)'s strengths and limiters.

Needs Analysis – What Is It?

A needs analysis is a process of deciding the physical and performance attributes necessary for an athlete to succeed in a particular sport. Another way to view the needs analysis is as a process that allows individuals to identify their strengths and limitations about a specific task or operation. An injury analysis of the mission and individual is performed as part of the process. This part of the needs analysis helps tailor the training toward preventing such injuries. Overall, the primary aim of a needs analysis is twofold [1, 2]. The first primary aim is to prevent and reduce injury, and the second is to enhance athletic performance.

To achieve these aims, the needs analysis process involves three steps [1, 2]. The first step is conducting the mission-oriented needs analysis; the second is the human

C. Myers, *The Human Weapon System*,
https://doi.org/10.1007/978-3-031-45061-7

weapon system needs analysis. The third and ultimate step is comparing the mission, and human weapon system needs analyses. The process includes the following steps:

Step #1: Organization-oriented needs analysis

(a) Commander's mission, vision, and intent
(b) Injury analysis of the organization

Step #2: Human weapon system needs analysis/athlete profile
Step #3: Comparative analysis of the mission-oriented and human weapon system needs analyses

The following sections cover the human performance optimization (HPO) program's needs analysis for Arctic operations. The HPO staff used real-world data and peer-review literature to conduct this analysis.

The Organization-Oriented Needs Analysis

Commander's Mission, Vision, and Intent

Mission

The 7th Infantry Company (Arctic) executes expeditionary operations within the Alaskan and Arctic Circle theaters, defends critical infrastructure in homeland defense, and on order is capable of decisively defeating any adversary in extremely cold weather, mountainous, high latitude, and high-altitude environments through large-scale combat operations.

Vision

Leveraging the strength of our people, this unit is ready to deploy, fight, and win decisively against any adversary anytime and anywhere in a joint, multi-domain environment across the spectrum of conflict and operations. Our organization is led by adaptive leaders of character who master the fundamentals, seek continuous growth, and execute the Commander's intent as part of a cohesive team.

Key Priorities

These key priorities are the operational approach to complete this unit's mission.

1. **Care.** This unit prioritizes care for soldiers, civilians, families, and veterans.
2. **Readiness.** This unit rapidly deploys from a capable power projection platform to fight and win in any environment, defend the homeland, or execute defense support of civil authorities.
3. **Leader development.** This unit heavily invests in its great asset, our people. This unit develops leaders of character who are resilient and thrive in complex and unpredictable environments.

4. **Teamwork.** This unit builds and sustains mutual trust and cohesive teams that share purpose and understanding.

Organizational Configuration

The 7th Infantry Company is comprised of 150 personnel. The professions include infantrymen (11B), medics (68W), supply (92Y), and communications (75B).

Training and Mission Execution

The unit has developed its own techniques, tactics, and procedures (TTPs) following the National Strategy for the Arctic Region. Mission execution occurs in the upper portions of Alaska and the Arctic Circle. Most missions are land based, but the unit must be prepared to work with the Marines and Navy for sea-oriented scenarios.

Critical general and specific tasks that the unit relies on to complete the mission are as follows:

1. Snow showing
2. Cross-country skiing
3. Snowmobiling
4. Tactical Combat Casualty Care (TCCC)
5. Move under fire with limited to no cover or concealment
6. Navigation
7. Voice communications
8. Employment of different weapons
9. Vehicle maintenance and repair

Training occurs several times per year. Training exercises can last from 6 h to several weeks in the field. While in garrison, the HPO team can expect to have up to 6 weeks of interaction with each HWS between training cycles.

Terrain and Weather

Arctic/Extreme Cold Weather Regions

Environments with ambient temperature at or below 0 °C are considered cold environments. Weather is considered extreme cold when the ambient temperature is at or below 0 °C, and temperatures drop below normal as wind speed increases. Four types of cold environments exist, which are as follows: polar, high mountain, glacial, and periglacial; the latter mainly represented by the Arctic and Antarctic regions. These regions receive less intensive solar radiation because the sun's light hits the earth at an oblique angle. As a result, the radiation spreads over a larger area and passes through the earth's atmosphere at longer distances. Thus the energy gets absorbed, scattered, and reflected. The present-day global distribution of these four environments is shown in Fig. 1.

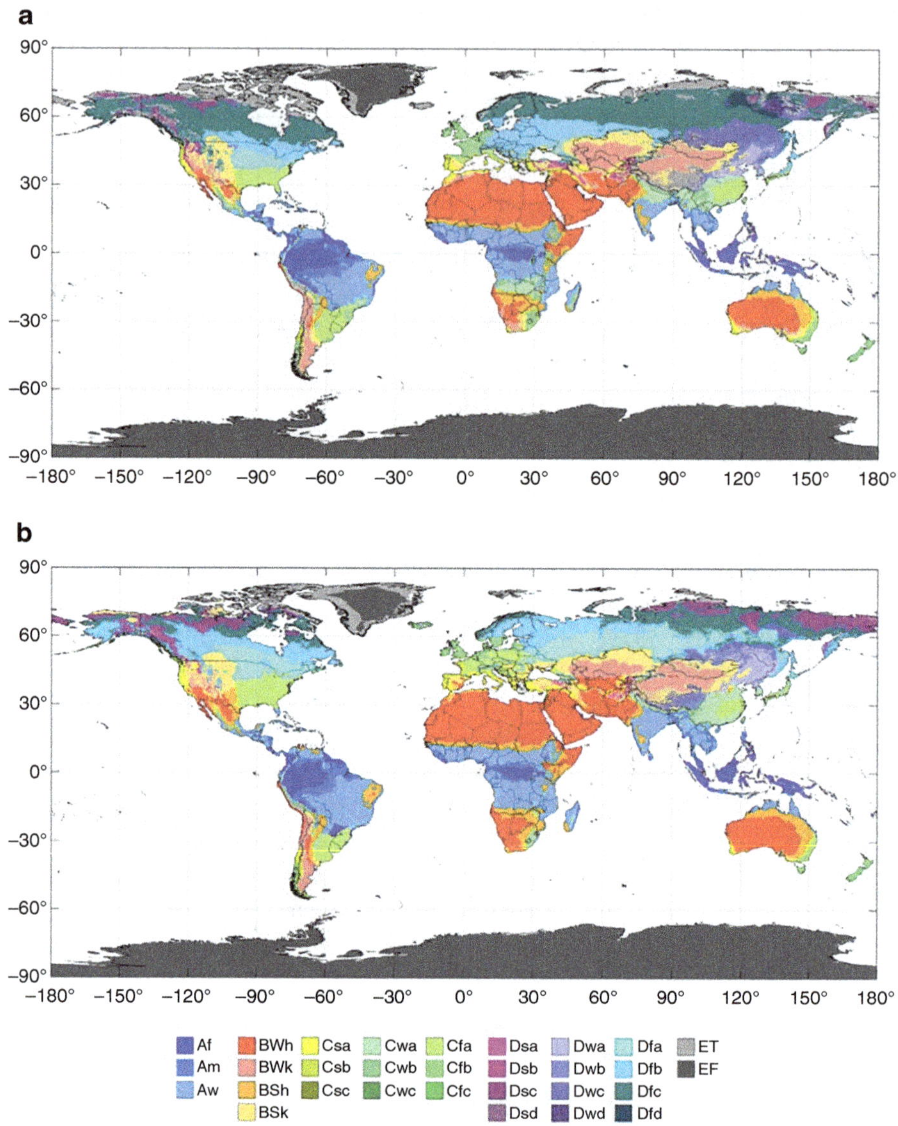

Fig. 1 The different polar regions of the earth [3]

The Arctic and Antarctic regions are north and south of their 60° latitude line. About 10% of the earth's land surface is covered by ice. Antarctica and Greenland account for 95% of Earth's ice content, while Canada and Russia account for the rest. Icy conditions can also be found in the mountain regions, such as the Himalayas, Andes, and Rocky Mountains. In Antarctica and Siberia, temperatures below −70 °C are observed in winter. The northern polar regions and Antarctica are arid

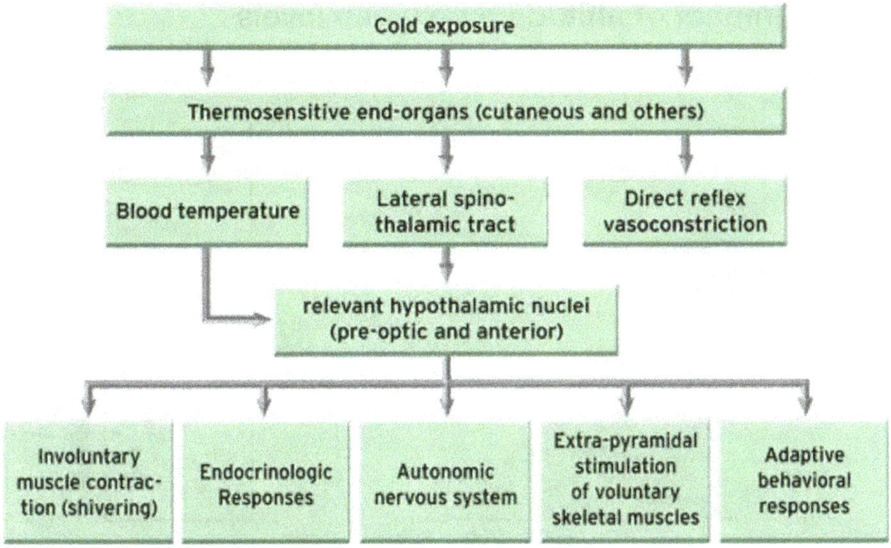

Fig. 2 Physiological pathways activated by cold exposure in humans

environments with relatively low amounts of precipitation (i.e., snow accumulation) located on the fringes of polar or glacial environments. While some enormous glaciers are found in polar environments, most of the world's actively moving glaciers are high in high mountain regions (alpine environment). The heavy winter snowfall in these areas provides the ice to feed the glaciers. Then, meltwater lubricates the glaciers in the summer, allowing them to move like giant conveyor belts down the valleys (Fig. 2).

High mountain areas experience frigid winters with heavy snow. Because of the high altitude, the temperature can drop far below −30 °C, and strong winds worsen the environmental conditions for humans. The extreme winter cold is replaced with warmer weather in the summer. At elevation, the partial pressure of oxygen drops (Fig. 3). This makes breathing difficult and reduces human performance until acclimation occurs. When the ambient temperature is below 0 °C, human physiological adaptations are minimal (Fig. 2), and, at elevation, the decreased partial pressure of oxygen causes decrements to these adaptations. The human body's heat-producing mechanisms are as follows: involuntary muscle contractions to produce heat, non-shivering thermogenesis by beta-adrenergic activation of brown-fat tissue – behavior, technological inventions, and logistical support play a central role in surviving in such a cold environment (Fig. 2).

Conclusions

1. The low temperatures and seasonal precipitation present a high risk for cold weather injuries and hypothermia. This risk increases at elevation.

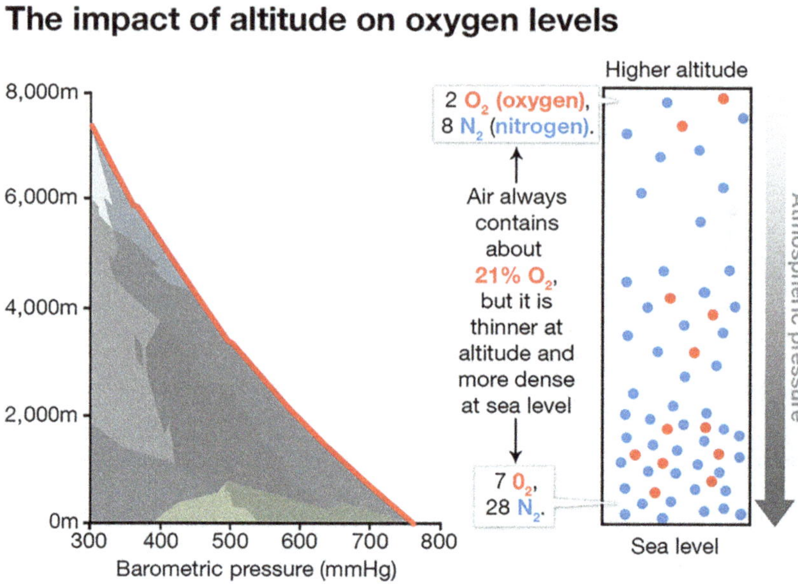

Fig. 3 This image depicts the impact of the altitude of oxygen. (www.theconversation.com)

2. Altitude sickness can be an issue. Decrements to muscular strength, muscular endurance, and aerobic endurance begin to decrease when one goes to elevations higher than usual. The human takes 16–21 to acclimate to altitude changes fully. Altitude sickness becomes an issue at or above 2500 m (8202 feet). The Human weapon systems experience increased resting heart and respiration rates at the mission elevation due to the slight drop in partial pressure of oxygen (PO_2) until acclimation occurs (Fig. 3).

Key Questions

1. How do you acclimate the human weapon systems to the temperature and weather conditions?
2. What clothing and personal protective equipment (PPE) are available to the human weapon systems for the warfare school and the exercise?

Physiology of Cold Exposure

Cold weather exposure increases the metabolic demand on the human body due to the thermal strain to maintain thermal homeostasis. As shown in Eq. 1, body temperature (BT) or heat storage (S) is the sum of all heat produced plus the heat transfer between the human body and the environment [4, 5]. The equation for S is as follows:

$$S = M - (\pm W_k) \pm E \pm R \pm C \pm K \left[W/m^2 \right] \tag{1}$$

whereas
S = heat storage
M = heat production
W_k = energy leaving the body (positive for concentric work and negative for eccentric work)
E = evaporation
R = radiation
C = convection
K = conduction
Units = Watts/meter^2

This equation depicts how to calculate heat storage for the human body (Fig. 4).

This equation is essential because the body reacts with its environment to maintain thermoregulation. Humans exchange heat with their external environment in four ways: evaporation, radiation, convection, and conduction. Figure 5 demonstrates the four avenues.

The primary heat loss mechanisms in dry air are conduction and convection – however, the exchange of heat increases when the body is wet. The human body loses heat up to 25× faster in water versus dry air. No matter the environment, heat will flow out when the external temperature is less than the BT. So, the lower the temperature, the more heat is lost. **This increases the thermal strain on the human body and increases metabolic demand.**

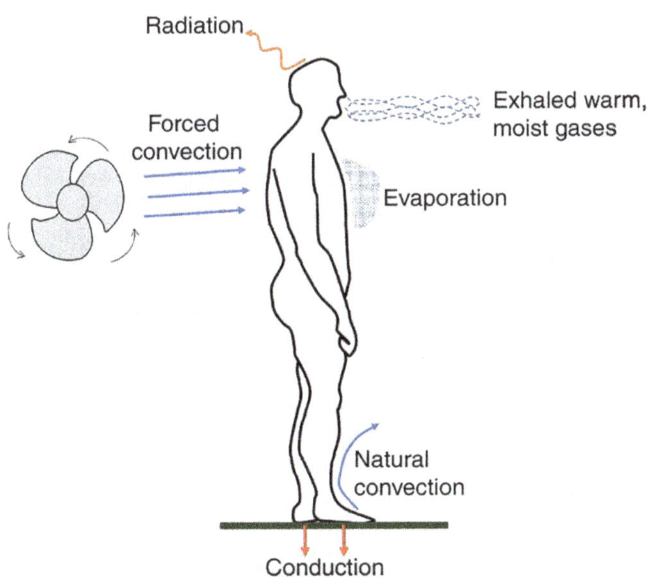

Fig. 4 How the human body exchanges heat with the external environment

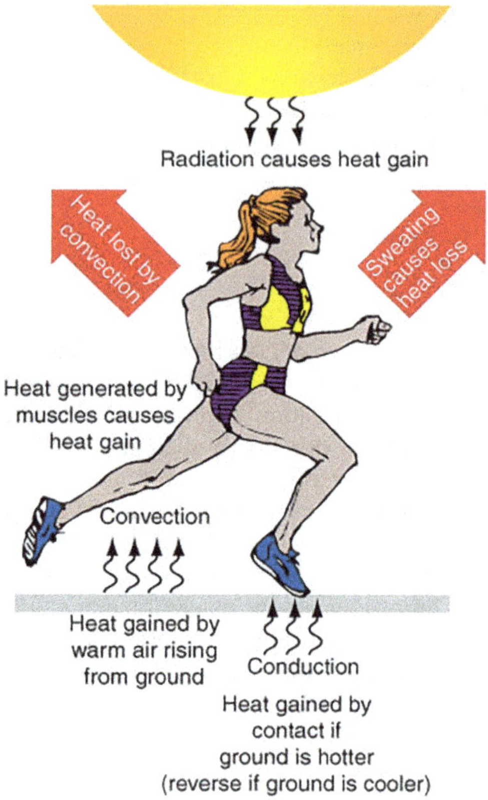

Fig. 5 The four avenues the body exchanges heat during activity

To help maintain thermal regulation, the human body has several responses. The first is vasoconstriction of the peripheral limbs. By restricting blood flow to the limbs, the body can limit heat loss to the environment and keep the core warm. Then, the body creates heat by contracting skeletal muscles.

About 70% of the heat generated by the human body, especially in colder temperatures, comes from muscular contractions [4, 5]. Therefore, physical activity (PA) creates more heat than shivering. Without PA, the body begins shivering in the torso region, spreading to the peripheral limbs. With PA, heat production can increase 2×–5× times, depending on the PA intensity [4, 5].

In colder conditions, increased muscular activity increases resting lactate levels [4, 5]. This is a strong indication of increased glycolytic activity to produce energy. Furthermore, research strongly suggests that glucose availability is a determining factor for maintaining heat capacity [6, 7]. However, it is still being determined if the limiting factors are blood or muscle glucose concentrations [6, 7].

The increased metabolic activity with increased vasoconstriction increases the cardiac output (Q).

Conclusions

1. The body loses heat via conduction, convection, evaporation, and radiation.
2. Skeletal muscle contractions generate above 70% of heat.
3. PA increases the amount of heat produced in the human body.

Key Questions

1. Will the PPE provided maintain the heat in the environment where the exercise takes place? Should active or passive heating be utilized?
2. What is the source available to provide external heat?
3. What is the caloric demand in this environment?
4. Will acclimation reduce the thermal strain on human weapon systems?
5. How does the increased cardiovascular load affect the human weapon system's performance?

Movement Modality and Physical Exertion

Typical cold operation exercise modalities are snowshoeing, cross-country skiing, and ruck marching in deep snow. In all three modalities, the exercise intensities can range from moderate aerobic to anaerobic. Comparing these intensities to the Dr. Andrew Coggan heart rate zones, the intensities range from Tempo/Zone 3 to $VO_{2\,max}$/Zone 5, as shown in Fig. 6. In the research conducted by Castellani et al. and Beals et al., ruck marching and cross-country skiing intensities primarily fell into low to mid-aerobic (Zone 2 and Zone 3).

Level	Name/purpose	% of threshold power	% of threshold HR	RPE	Time
1	Active recovery	≤55%	≤68%	<2	1.5hours
2	Endurance	56-75%	69-83%	2-3	2.5 hours to 14 days
3	Tempo	76-90%	84-94%	3-4	30min to 8 hours
4	**Lactate threshold**	**91-105%**	**95-105%**	**4-5**	**10 - 60 min.**
5	**VO_2max**	**106-120%**	**>106%**	**6-7**	**3 - 8 min.**
6	**Anaerobic capacity**	**121-150%**	**N/a**	**>7**	**30 sec. - 2 min.**
7	Neuromuscular power	N/a	N/a	(maximal)	5 - 15 sec.

Fig. 6 Dr. Andrew Coggan heart rate zones. (www.peakscoachinggroup.com)

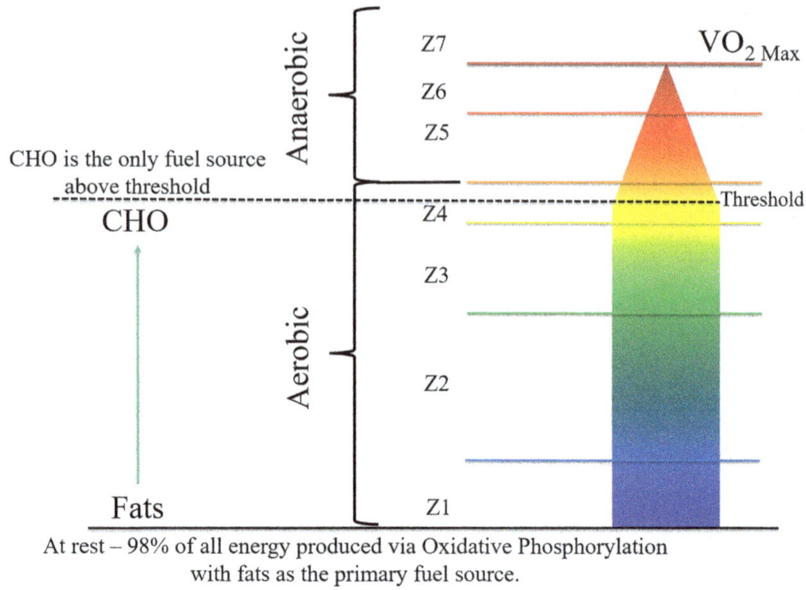

Fig. 7 $VO_{2\,max}$ concept with Dr. Coggan heart rate zones

The exercise duration and intensities greatly influence the fuel source for energy production. The lower the exercise intensity, the more reliant the body is on fats as the fuel source. However, the closer you reach the threshold, the more reliant the body is on carbohydrates [8], as shown in Fig. 7.

Coupling the exercise intensities and the effects of freezing weather on carbohydrate metabolism, the human weapon systems will heavily rely on carbohydrates as the primary fuel source for this mission set.

Cold Weather Operations and Nutrition

The increased metabolic demand of muscular contractions to maintain thermal balance via the thermal strain causes higher than normal Total Daily Energy Expenditure (TDEE). Energy deficiency is a problem in these types of operations. Energy deficiency is reported to range from 4200 to 7100 kcal/day [9, 10]. In the peer-reviewed journal, "International Journal of Sports Nutrition and Metabolism" (impact factor 3.884), a study by [11] in 2019 investigated energy deficiency in 10 male Naval Special Warfare (NSW) Sea, Air, and Land (SEAL) qualification students during the 3-day mountain warfare/cold weather (MWCW) exercise. The students performed river crossing (RC) and mountain patrol (MP) operations during the exercise. During the mountain patrol operation, the students covered 6.1 km, burned an estimated 5402 ± 922 kcal, and ate an estimated 2289 ± 680 kcal. This left a caloric deficit of 3112 ± 1420 kcal. Another study in the peer-reviewed journal *Extreme Physiology and Medicine* (impact factor = 2.8) by Castellani et al., 2017 followed 18 male Norwegian soldiers through similar conditions and terrain described earlier in the analysis. The study demonstrated an estimated energy expenditure of 6821 ± 578 kcal/day while in-taking an estimated 3465 ± 622 kcal/day. This left an estimated caloric deficit of 3357 ± kcal/day.

In both studies, the soldiers were given military rations. The Beals et al. and Catellani et al. do not say if supplements were taken. According to Defence Logistical Agency (DLA), the agency provides 12 menus of "Meal, Cold Weather/Long-Range Patrol (MCW/LRP)" meals for cold weather operations (www.dla.mil). The average MCW/LRP contains 1540 kcal (15% protein, 35% fat, and 50% carbohydrate). It is suggested that three menu bags per person per day provides 4500 kcal necessary to replenish energy loss from exertion in extreme cold. However, the recommended number of calories is a minimum of 2321 kcal short of the expected energy expenditure range of 6821 kcal/day as laid out by Castellani et al. since this study was conducted in similar terrain, conditions, and modality; this needs analysis is investigating. In addition, these estimations are global, meaning there will be variations from person to person. However, knowing these metrics will help plan and provide a majority solution for human weapon systems.

The MCW/LRP, on average, has 15% protein, 35% fat, and 50% carbohydrate (www.dla.mil). The macronutrient percentages for these meals do not fit the suggested requirements by recent research. The following will look at each macronutrient.

Carbohydrates

As stated earlier, the body uses muscle contractions to generate most of its body heat. At rest, shivering is the primary modality to increase BT. This action increases resting blood lactate levels, indicating the body uses anaerobic glycolysis as the metabolic pathway to produce energy. Carbohydrates are the only fuel source for this pathway [1, 8, 12, 13]. Vallerand and Jacobs demonstrated that shivering increases carbohydrate utilization 7× more than fat oxidation [14]. To fuel shivering at rest, carbohydrate is 51% of the total energy expenditure compared to only 18% at an average temperature. In this context, the MCW/LRP provides the right proportion of carbohydrates at rest. However, some emerging research suggests that fat oxidation increases in cold temperatures compared to normal and warmer temperatures [15]. However, the temperatures in the laboratory research did not reach the temperatures or mimic the exercise modality of typical cold weather operations. The critical difference in stating that the MCW/LRP does not provide the right carbohydrate proportion is the inclusion of exercise modality.

The exercise modalities the human weapon systems will be exposed to (cross-country skiing and ruck marching) are higher than average intensities. Even though these intensities are submaximal, they will be very close to not at the threshold. This can be said by looking at the provided research. In the study by Beal et al. investigating mountain training with the NSW SEAL qualification students, the students spent 41% of their time at or above the threshold (Beal et al. 2019) during the mountain patrol. At this intensity, carbohydrates are the primary fuel source to produce energy [1, 8, 12, 13]. In the research by Castellani et al., which mimics the conditions this needs analysis is investigating, the Norwegian soldiers had heart rates in the upper 130s–upper 140s beats per minute (BPM) during movements and cross-country skiing. For a trained athlete, these intensities are upper endurance, where fats would be the predominant fuel source; the typical threshold heart rate is in the low to mid-160s. However, our human weapon systems have lower threshold heart rates. Our human weapon system's average threshold heart rate appears in the low-to-mid-150s BPM. The lower threshold range and lack of specific training for cross-country skiing and cold weather mountain operations increase the individual intensity of our human weapon systems, thus increasing the reliance on carbohydrates as the fuel source. Putting all this information together strongly indicates that the carbohydrate proportion needs to be greater than 50% of the daily caloric intake, as suggested by DLA. Therefore, each human weapon system should target about 60% of Total Daily Energy Intake (TDEI) from carbohydrates.

Proteins

The MCW/LRP hits the right proportion for proteins. The minimum recommended dietary intake (RDI) for protein intake is 0.8–1.0 g/kg of body weight (BW). However, in situations where caloric deficits are seen, research suggests the RDI for protein to be 1.2–1.7 g/kg of body weight [16]. At 15% protein of 1500 kcal meal, the MCW/LRP provides about 56 g of protein. The average human weapon system is 86 kg (190 lbs.). Therefore, MCW/LRP meals would provide 1.5 g/kg of body weight.

This estimate is based on the average weight of every human weapon system. Each athlete will have individual requirements. The protein from the MCW/LRP meals is a good starting point. However, it is suggested that the protein proportion be closer to 1.7 g/kg of body weight to help maintain muscle mass during the caloric restrictions and to help promote muscle recovery.

Fats

With the suggested increase in carbohydrate ingestion, less reliance on fats is necessary. It may be prudent to consider supplementing some fats to address the caloric deficit; however, the focus should be on carbohydrates and protein. The MCW/LRP meals appear to provide enough fats for daily energy requirements if only carbohydrate and protein supplementation were utilized to close the caloric deficiency gap.

Conclusions

1. The estimated Total Daily Energy Expenditure (TDEE) for the human weapon system ranges from 5402 ± 922 kcal/day to 6821 ± 578 kcal/day. This range is highly dependent on temperature conditions and exercise modality. The two studies in this section state that the average TDEI ranges from 2289 ± 680 kcal/day to 3465 ± 622 kcal/day. The DLA suggests that caloric intake is 4500 kcal/day. If eating only as per the DLA MCW/LRP meal recommended meal allocation, the caloric deficit could range from 902 kcal/day to 2321 kcal/day.

2. The nutritional strategy should address an emphasis on carbohydrate and protein intake. Based on the research discussed thus far, the body becomes highly reliant on carbohydrates as a fuel source. However, the proportion of carbohydrates in the MCW/LRP meals is low. Therefore, when trying to make up for the caloric deficit, most of the calories should come from carbohydrates. Additionally, protein supplementation should **help** mitigate muscle atrophy and increase recovery.

Key Questions

1. What is the best way to close the caloric deficit? Should extra MCW/LRP meals be provided to each human weapon system? Are nuts and small-sucking candies a more significant part of the equation?
2. Can commercialized carbohydrate and protein supplements be utilized in these conditions?
3. Should the human weapon systems follow a fueling plan during the operations to ensure the proper carbohydrate ingestion rate? What would the ingestion rate be? Is the rate similar to that used for day-to-day endurance training (~60 g/h), or is it higher due to increased reliance on carbohydrates as the fuel source? Is the 1.0 glucose:0.8 fructose carbohydrate ratio still optimal for these conditions?
4. Should whey protein be the only protein supplementation, or should casein be considered?

Hydration in Extreme Cold Weather

Hydration does not mean drinking water. This term encompasses replacing water and minerals the body loses during perspiration. The process of perspiration (i.e., sweating) is the release of a water solution from the eccrine glands in the skin to help regulate body temperature. As Fig. 8 details, sweat contains much more than water.

Each component lost during perspiration is essential to replace because each element is involved with critical biological processes such as nerve conduction, cardiovascular function, and energy products.

Sweating rates in cold weather are like warm conditions, especially with exercise. As stated earlier, most of the body's heat is through muscular contractions. Therefore, the body will produce more heat through exercise than shivering alone. This means the body will sweat when the core temperature increases above 97.6 F/36.6 °C.

Each athlete sweat at different rates and solute concentrations [17]. When fully loaded, the human weapon system can sweat like an American football player in normal temperature conditions. Additionally, the average time for a military operation is similar to that of a cycling or IRONMAN event. In both cases, the estimated sweat rate for football players and cyclists is about 1.5 L/h, as shown in Fig. 9. It is important to note that very little published data on military sweat rates exists.

Fig. 8 Sweat contents besides water. (Baker and Wolfe [16])

	Concentration
Sodium	10-90 mmol/L
Chloride	10-90 mmol/L
Lactate	5-40 mmol/L
Urea	4-12 mmol/L
Potassium	2-8 mmol/L
Ammonia	1-8 mmol/L
Others (e.g., bicarbonate, calcium, magnesium, glucose, amino acids, iron, copper, zinc)	< 1 mmol/L each

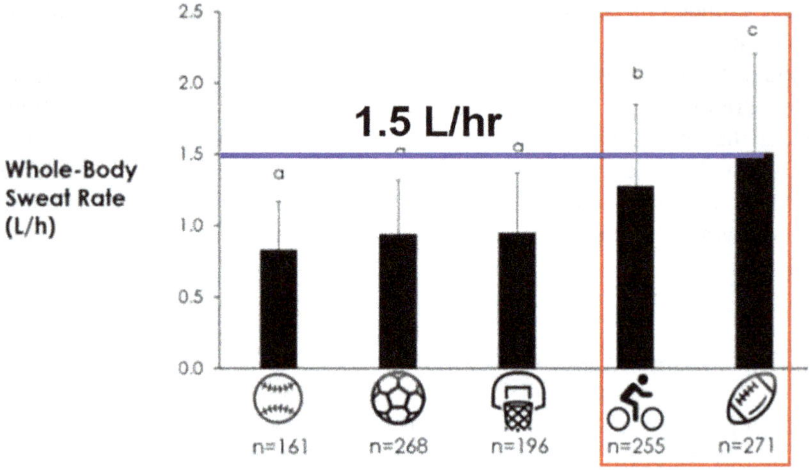

Fig. 9 Normative sweating data for several different sports in average temperatures. [18]

During cold weather operations, the human weapon system needs to target 1.5 L/h and ensure the micronutrients of sodium, magnesium, and calcium (at a minimum) are part of the hydration solution.

The reason for addressing hydration is essential, if not more important than nutrition. Maintaining water balance is critical. First, if an athlete has a 2% drop in water weight, there is typically a 5% drop in performance. This decrement in performance is not linear; it is exponential. So, the more water lost, the worse the drops in performance are.

Additionally, more than drinking water is needed to maintain hydration. An athlete can drink too much water and insufficient electrolytes and minerals to cause hyponatremia. Hyponatremia is when sodium levels in the blood are lower than usual. This condition can cause muscle cramps, weakness, nausea, lethargy, headaches, mental confusion, and seizures. All dehydration and hyponatremic conditions can be mitigated with a simple hydration strategy. Some basic hydration guidelines are as follows (Baker 2021):

- **Prevent > 2% drop in BW from water loss**
- **Avoid overdrinking (increase in BW)**
  - This could lead to hyponatremia.
- **Consume sodium with fluid if significant sweat sodium volume losses (2–3 g) occur:**
  - High sweat rates (1.5 L/h)
  - "Salty Sweat" (> 60 mmol/L)

A sports drink typically has the correct balance of water, electrolytes, and minerals. Drinking at a rate of 1.5 L/h with a mix of 40–50% water and 50–60% sports drink is a solid starting point for planning.

Conclusions

1. Replacing the water and components of sweat from perspiration is essential to basic biological functions.
2. The human weapon system should use the sweat loss rate of 1.5 L/h to plan a hydration strategy.
3. Any hydration strategy should utilize a mixture of sports drinks and water.

Key Questions

1. What is the best commercial sports drink product in these conditions?
2. Can a tablet form reduce the amount of liquid carried?
3. How to keep liquid from freezing in these conditions?

Sleep and Recovery in Cold Weather Operations

The quality and amount of sleep during sustained operations take much work to quantify. However, more than sleep quality and quantity are needed to maintain high human performance. Based on subjective reports the HPO team has received from early 2021 field operations, sleep typically averages 3–5 h with poor quality. Additionally, Arctic Infantrymen must stay at the station from 12 h to 16 h. With the elements of reduced sleep and longer time on station, several issues will occur:

1. Reduced cognitive ability ([19, 20]; Patrick et al. 2017; [21–23])
2. Reduced hormonal levels (Gopalakrishnam et al. 2004; [24])
3. Disruption in circadian rhythms (Gopalakrishnam et al. 2004; [24], [25])
4. Disruption and reduced protein synthesis for muscle recovery ([24, 26–28])

Research shows that after acute reductions in sleep (<5 days), mental acuity drops by over 50%, and protein synthesis drops by 30%. Chronic sleep debt (>7 days) greatly exacerbates these four issues. Similar studies with sleep debt over 9 days show almost double mental acuity and protein synthesis decrements. In addition, the recovery time was nearly triple (nearly 120 h of total sleep) over a 5–7 period of time.

Furthermore, sleep deprivation in average temperature does not alter the core temperature with exercise in colder temperatures [29]. However, sleep deprivation is shown to cause decrements in the body's ability to coordinate fluctuations in skin temperature. This suggests that sleep deprivation alters the body's ability to mediate skin blood flow and baroreceptor-mediated cardiovascular regulation (Remeijn et al. 2012). So, this means sleep deprivation significantly decreases the body's thermoregulatory mechanism. One way to combat this decrement is to maintain the ability to exercise.

Conclusions

1. The average Arctic Infantryman's amount of sleep in the field is typically 3–5 h, with the quality of sleep rated as poor.
2. Sleep deprivation causes significant drops in mental acuity and causes substantial disruptions to thermoregulation, circadian rhythms, and hormonal responses.

Key Questions

1. What is the appropriate work-to-rest ratio?
2. How can the limited amount of sleep quality be improved during Observation Post (OP) operations?

Organization Injury Analysis

The following details a list of possible injuries from cold weather exposure and exercise modalities (Table 1).

Known Injury Statistics

This section is based on the information gleaned from an exercise in the Arctic Circle. During the brigade level, 9-day exercise, temperature ranged from −11 F to −55 F. During the exercise, there were 36 cases of frostbite (34 superficial, 2 deep), 32 cases of chilblains, and no hyperthermia cases.

Frostbite

Frostbite Contributing Factors (Table 2)

- Previous history of cold weather injuries
- Location of mechanism of injury
- Amount of time lived in Alaska prior to the exercise
- Type of cold weather training prior to the exercise
- Tobacco use

Table 1 This table depicts injury analysis

Activity	
Cross-country skiing	
Ruck marching	
Run	
Physiological aspect	
Metabolic	
Anaerobic (phosphagen & glycolytic) – 5–10% of total time	
Aerobic (oxidative phosphorylation) – 90–95% of the total time	
Range of motion (ROM)	
Shoulders and back	Hips
Ankles	Knees
Flexibility	
Shoulders and back	Hips
Ankles	Knees
Common injuries	
Cold weather injuries	Dehydration
Ankle sprains and strains	Medial Cruciate Ligament (MCL) sprains and tears
Anterior Cruciate Ligament (ACL) sprains and tears	Ulnar collateral ligament of the thumb strains
Lower back strains	Fatigue-related injuries

Table 2 This table lists the degrees of frostbite injury and symptoms

Degree of frostbite	Appearance after rewarming
First-degree frostbite	Numb central white plague surrounded by erythema, but no blistering
Second-degree frostbite	Blister formation surrounded by erythema and edema. Blisters filled with clear or milky fluid in the first 24 h
Third-degree frostbite	Death of skin and subcutaneous tissues forming hemorrhagic blisters in black eschar 2–3 weeks later
Fourth-degree frostbite	Tissue necrosis, gangrene, and eventually full-thickness tissue loss; initially, the body part is hard, cold, white, and numb post-rewarming

- Amount of sleep prior to cold weather injury (CWI)
- Diet prior to CWI
- Type of gloves worn
- Type of boots worn
- Vapor Barrier (VB) boot valves open, closed, and serviceable
- Commercially purchased clothing, not using military issued clothing (Figs. 10, 11, 12, 13, 14, and 15)

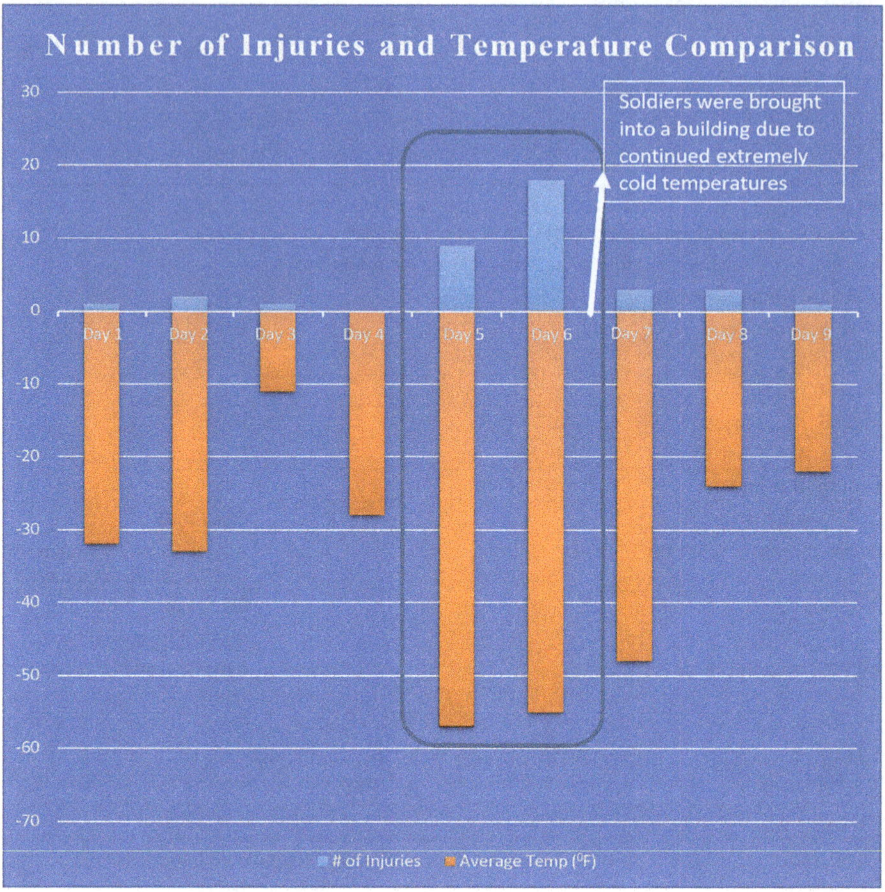

Fig. 10 This figure depicts the number of injuries compared to temperature exposure

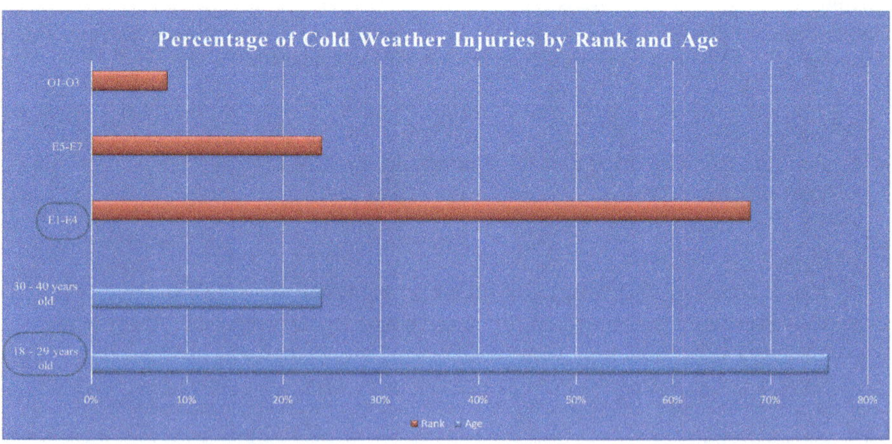

Fig. 11 Depicts all cold weather injuries by rank and age

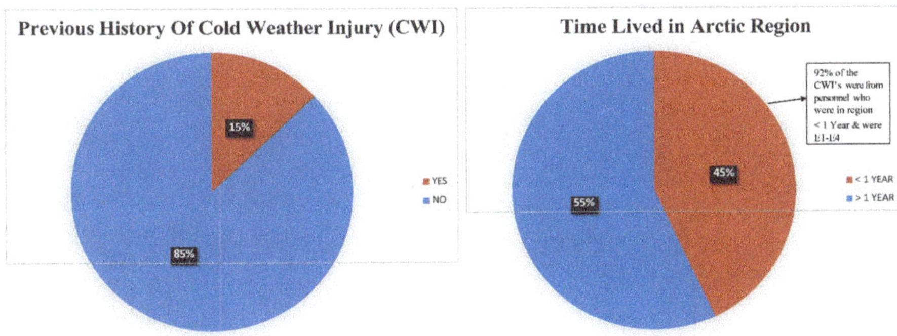

Fig. 12 Depicts the percent of CWIs and time in Arctic region of all personnel participating in the 9-day exercise

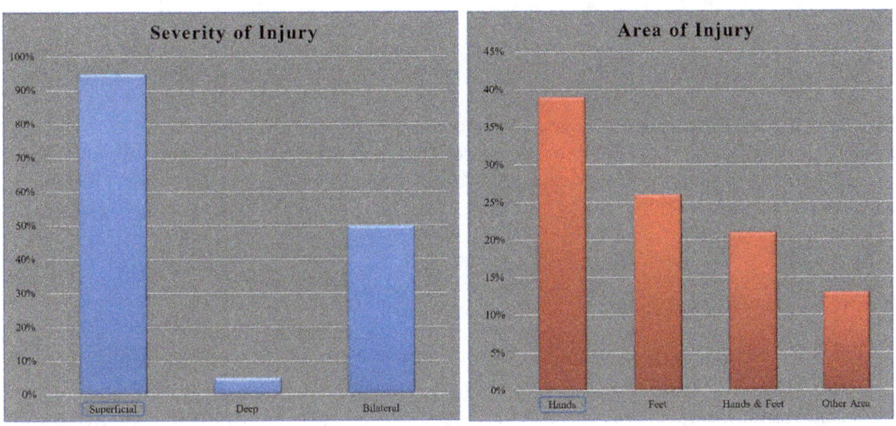

Fig. 13 This figure depicts the severity and location of all reported CWIs during the 9-day exercise

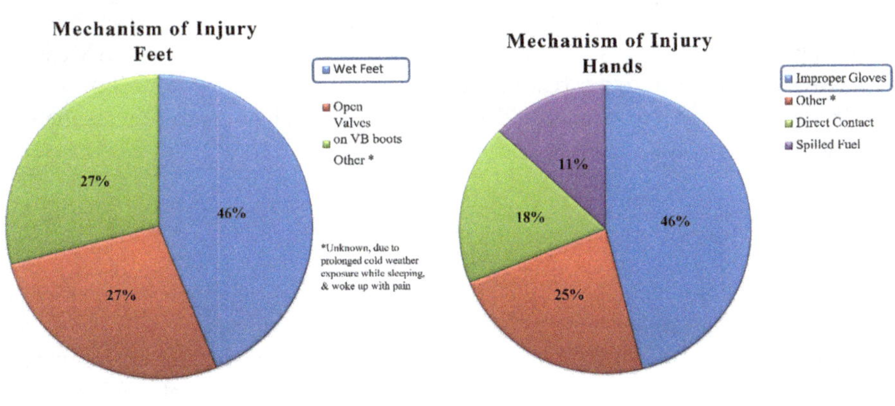

Fig. 14 This figure depicts the mechanism for CWIs of hands and feet

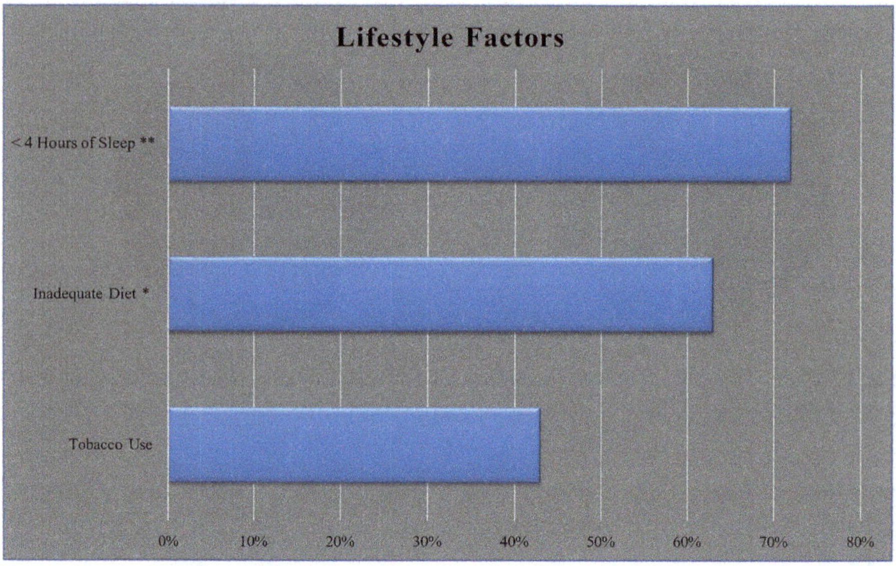

Fig. 15 This figure depicts the lifestyle factors that contributed to all reported CWIs. Average sleep duration 2–3 h due to high op-tempo and/or being too cold to sleep. Frozen water prevented soldiers from reconstituting their Meal Ready to Eat (MREs) for consumption. Others did not eat secondary meal to lack of time due to op-tempo of exercise, and others did not want to eat because of the cold

Known Frostbite Causes

1. Lack of adequate cold weather training before major exercises (human performance research for military operations in extreme cold environments).
2. Acute cold weather injuries are typically underreported when the average temperature is between 0 F and +20 F. COVID-19 significantly limited cold weather training opportunities.
3. Cold weather injury pentad = lack of sleep, inadequate nutrition, tobacco use, improper equipment (improper gloves, VB boot valves), and wet extremities.
4. Younger, junior soldiers with less than 1 year in Arctic/extreme cold weather environments are at higher risk for injury.
5. 100% of cases where soldiers spilled Jet Propellant 8 (JP-8) onto skin resulted in frostbite.

Conclusions

1. Lower body injuries make up the predominance of injuries. Increases in mobility, strength, and endurance can help mitigate these injuries.
2. Improve leader education on modifying op-tempo in cold weather environments to allow for "sleep breaks" and adequate nutrition to prevent cold weather injuries.

3. Incorporate outdoor cold weather exercises into quarterly training plans that provide a crawl, walk, and run training experience.
4. Empower younger, junior soldiers to teach portions of cold weather training.
5. Thorough leader checks to ensure soldiers have proper gloves, VB boot valves are closed, and feet are dry.
6. Continue to push for tobacco cessation (including leaders) – minimize chewing tobacco during field events.

Key Questions

1. N/A.

Arctic/Extreme Cold Weather Clothing

Typically, Arctic/extreme cold weather clothing is designed to work in static and active employment from temperature ranges down to −40 °F for 1 h with a light metabolic output threshold, −100 °F for 4 h with a light metabolic output objective of −70 °F. Additionally, the clothing must be scalable. Currently, Air Force Special Warfare (AFSPECWAR) utilizes Generation III Extended Cold Weather Clothing System (GEN III ECWCS). This system was initially issued to soldiers in Afghanistan in 2007. The materials used in this system offer a greater range of breathability and environmental protection and greater versatility compared to previous editions.

"GEN III ECWCS is a 12-piece kit (seven-level) that enables users to utilize seven different layers (weight = 12.82 lbs./5.82 kg), depending on the mission and environment. The system functions through insulation, which resists the transmission of heat, traps air, and wicks moisture away from the body; layering, which increases air space and allows easy adjustment to a user's activity level; and ventilation, which allows moisture to escape. By mixing and matching Gen III ECWCS components, Soldiers can protect themselves from temperatures ranging from 40 degrees Fahrenheit to minus 60 degrees" (https://www.military.com/equipment/extended-climate-warfighter-clothing-system-gen-iii). The seven-level system consists of the following items (https://www.military.com/equipment/extended-climate-warfighter-clothing-system-gen-iii):

1. Level I: Lightweight Undershirt and Drawers
   - Base layer: Worn next to skin by itself or in conjunction with other levels for added insulation and to aid in transferring moisture.
   - The GEN III lightweight undershirt and drawers are designed to wick moisture and to provide protection from the elements.
   - Constructed with Polartec Power Dry Silk weight material, this layer is highly breathable, wicks moisture away from the skin, and dries fast, providing evaporative cooling in warmer weather and insulating in cool weather – all with less weight and bulk than previous systems.

2. Level II: Mid-Weight Shirt and Drawers
    - Base layer: Worn next to skin by itself or in conjunction with other levels for added insulation and to aid in the transfer of moisture.
    - The GEN III Mid-Weight Shirt and Drawers provide light insulation in mild climates and serve as a base layer in cold climates.
    - Constructed with Polartec Power Dry grid material, this layer provides extra warmth in cooler conditions but still wicks moisture away and dries fast. The material offers stretch for increased comfort and is lighter with less bulk than previous systems.
3. Level III: High-Loft Fleece Jacket
    - Primary insulation layer: Worn underneath shell layers or worn as an outer garment in cool conditions.
    - The GEN III High-Loft Fleece Jacket is the primary insulation layer for use in moderate to cold conditions.
    - Constructed with Polartec Thermal Pro® material, the insulation on the level III jacket creates air pockets that trap air and retain body heat providing outstanding warmth without weight. With excellent breathability, the Level III jacket dries quickly with increased warmth while maintaining lower bulk and increased durability than previous systems.
4. Level IV: Wind Jacket
    - Shell layer: Designed to be worn with base and insulative levels in transitional environments to provide wind and sand protection.
    - The GEN III Wind Jacket is designed to act as a low-volume shell layer that optimizes the moisture-wicking properties of other GEN III insulative and next-to-skin layers. It is also designed to be worn under body armor.
    - Constructed from stretchable nylon with a water-resistant finish, this full-zippered jacket provides wind and sand protection with low weight and bulk.
5. Level V: Soft Shell Cold Weather Jacket and Trousers
    - Shell layer: Designed to be worn with base and insulation levels for use in moderate to cold conditions.
    - The GEN III Soft Shell Cold Weather Jacket and Trousers are designed for use in moderate to cold weather conditions as a soft shell combined with other base and insulative layers.
    - Constructed with stretchable and breathable water-resistant materials, the jacket and trousers provide lightweight (LW), low bulk, and extreme comfort for movement.
6. Level VI: Extreme Wet/Cold Weather Jacket and Trousers
    - Shell layer: A waterproof barrier designed to be worn over other levels in moderate to cold wet conditions alternating between freezing and thawing.
    - The GEN III Extreme Wet/Cold Weather Jacket and Trousers are designed for use in moderate to cold wet conditions as a hard-shell layer combined with other base and insulative layers.

- Constructed with two-layer GORE-TEX and seam-sealed throughout, the jacket and trousers provide an outstanding lightweight, completely waterproof, windproof, and breathable level of protection against the elements with 50% less bulk than previous systems. Level VI also incorporates near-infrared signature reduction technology further enhancing soldier survivability.

7. Level VII: Extreme Cold Weather Parka and Trousers
    - Soft-shell insulative layer: The outermost level of protection in the system designed for use during static operations in extreme cold and dry conditions.
    - The GEN III Extreme Cold Weather Parka and Trousers are designed for use during static operations in extreme cold, dry conditions.
    - Constructed with an outer shell fabric that has a water-resistant finish and with PrimaLoft Sport thermal bonded high-loft insulation, Level VII is highly durable and breathable. The Parka and Trousers maintain warmth even when wet and provide protection in extreme cold conditions during static operations.

In the mid-2010s, US Army Natick began to look at the Military Alpine Recce System (MARS) to replace the GEN III ECWCS. The MARS garments are developed by Patagonia Inc. The major difference with the MARS is it has no more than three layers. These layers consist of a base layer, cross layer, and outerwear (active or static). The MARS system is as follows:

MARS Active-Base Layers The MARS active-base layers are fabricated from a fire-resistant blended wool material designed for wear next to the warfighter's skin by itself or under other layers. There are two fabric weights: lightweight (LW) and mid-weight (MW). LW wool blend garments include a short-sleeve t-shirt, a long-sleeve t-shirt, and boxer briefs. MW wool blend garments include a balaclava, a zip neck long-sleeve thermal undershirt, and thermal bottoms (Fig. 16).

MARS Active Cross Layers The MARS active cross layers are fabricated from several synthetic materials designed for permeability (i.e., breathability), stretching, and durability (Fig. 17). Some of the garments also include durable water-resistant (DWR) characteristics.

MARS Active Outerwear Layers The MARS active outerwear layers are designed for pairing with next-to-the-skin layers for intense activity when warmth and water repellency are needed (Fig. 18). The active outerwear layers, similar to the cross layers, are fabricated from several synthetic materials designed for permeability (i.e. breathability), stretching, and durability. However, in order to provide greater environmental protection (i.e., wind and water resistance), the outer layers have less permeability and greater water repellency than the cross layers.

MARS Static Outerwear Layers The MARS static outerwear layers are designed for pairing with next-to-the skin layers and are intended to provide inclement weather protection during low output or static activities when warmth is needed (Fig. 19). The static outerwear layers are fabricated from several synthetic materials designed for durability and water resistance.

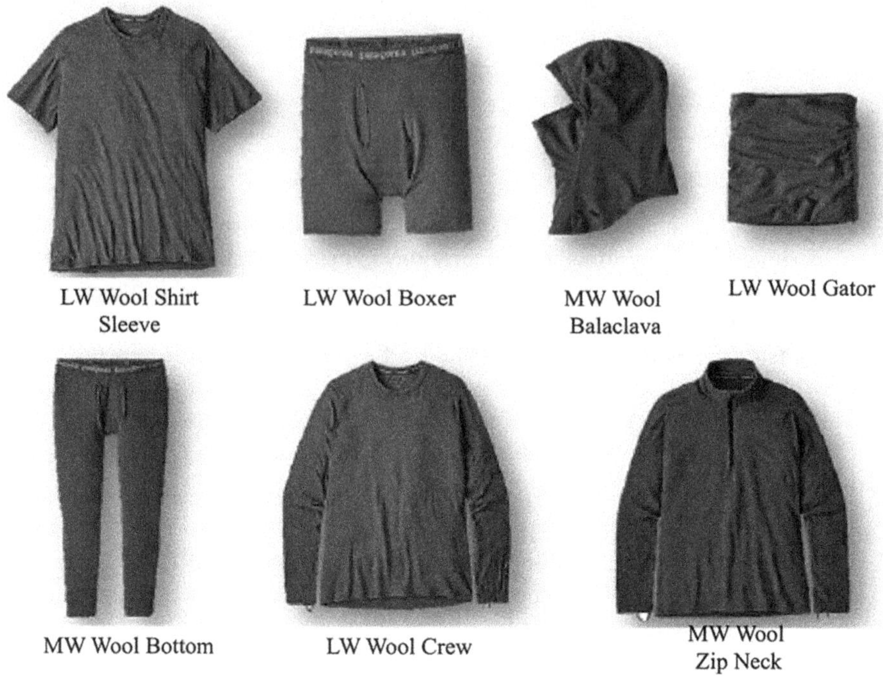

Fig. 16 Components of the MARS active-base layer

Fig. 17 Components of the MARS active cross layers

Fig. 18 Components of the MARS active outerwear

Fig. 19 The components of the MARS static outerwear

Given the options and complexity of the GEN III ECWCS, the Arctic Infantryman community continues to look at other Commercial Off-The-Shelf Sensors (COTS) options. The limitation of utilizing COTS options is due to the lack of testing and not meeting Department of Defense (DoD) garment requirements for this environment. The end user may not know if the COTS garment was properly tested. Usually, a manufacturer does not provide research data for a particular garment even when requested. Additionally, the DoD has set guidance that Arctic garments need to meet. Many times, COTS garments labeled at "extreme cold" do not meet these DoD requirements. **Not wearing DoD-approved extreme cold weather garments was one of the leading causes of CWIs in the 9-day exercise previously discussed**. Arctic Infantryman units and Arctic Infantrymen should be very cautious when purchasing and issuing nonissue extreme cold weather clothing.

The Human Weapon System–Oriented Needs Analysis

What Is the Human Weapon System–Oriented Needs Analysis

With the previous section looking at the mission demands, the next step in the needs analysis process is to look at the human weapon system itself [2]. The human weapon system profile can have the following list:

1. Goal setting
2. Athlete anthropometrics
3. Experience and competition level
4. Medical history
5. Training availability
6. Available equipment
7. Identify the athlete's strengths and limiters

Goal setting is arguably the essential part of the human weapon system profile – goals set the stage for expectations. Setting goals can be a difficult and daunting process. One way to make it easier is to create goals that are achievable, realistic, and tangible [2]. These three goal-setting criteria make achievable goals. The last thing a human weapon system or HPO personnel would like to do is create an unattainable goal. A challenging goal can be achieved with hard work. Setting the right goal(s) can set the stage for mission success and supply guidance on how the training program should be structured [2].

Gathering the human weapon system's anthropometrics provides insight into the athlete's body composition. These metrics include height, weight, and body fat percentage. This information indicates the athlete's physical fitness level and relative power for performance metrics [1, 2].

The next component in building the athlete profile is understanding the athlete's experience and competition level. With this information, one can begin to answer some of the following questions:

1. How long has the human weapon system been participating in the career field?
2. Does the human weapon system have the necessary skills and experience to reach the current goals?

With this information, one can begin to build the picture of the human weapon system. In addition, understanding the athlete's personal and career field-level experiences lays the foundation for the types of skills and intensity that the training program should incorporate [1, 2].

The next step is gathering the human weapon system's medical history [2]. This history provides the athlete or coach with valuable information that might supply insights into limitations to training. Also, this information provides insight into possible predisposing factors for injury. Finally, the medical history should be compared to the injury analysis performed in the mission-oriented needs analysis. This comparison provides the information necessary to see if the human weapon system is prone to or overly prone to the common injuries within the mission profile.

Training and equipment availability are usually something taken for granted. Attempting to create a training plan where the human weapon system needs to train for 20 h per week but only has 10 h a week to train is impossible. Also, ensuring the human weapon system has the proper training equipment sets him up for success. Knowing the available training time and equipment helps provide information on the type of training strategies that could aid the human weapon system in achieving its goals.

The second most crucial aspect of the athlete-oriented needs analysis is understanding the human weapon system's key limiters and strengths. This section is just as important as goal setting, but it must be performed last. Identifying a human weapon system's key strengths and limiters can come from two different avenues: either through objective or subjective means. What is meant by this is that the human weapon system can perform a self-analysis and identify its strengths and limitations. Another way to do this is by analyzing the human weapon system's historical data. Some data that can be examined are past workouts, field tests, and competition performances. However, the optimal way to identify these aspects of the human weapon system profile is through lab or field testing.

The information gleaned from the mission and human weapon system needs analyses provides the framework to create a personalized training program. However, the needs analysis process is incomplete until these two steps are compared, which occurs in the third phase of this process, termed the "comparative analysis" (Haff and Triplett 2016; [2]).

****Note. This section should have several HWSprofiles as identified in Step #1. However, the Arctic Infantryman profile is included in this example.**

The Human Weapon System/Arctic Infantryman Profile

Demographics

The average Arctic Infantryman is male with height ranging from 68 to 71 inches (175–181 cm) with an average weight of 190 lbs. (86.3 kg). The average Date of Birth (DOB) is 1988 ± 6.2 years. The average time in service is 10.7 ± 6.2 years.

The Arctic Infantryman is trained in small unit tactics and extreme cold warfare. In addition, some individuals have advanced training in survival, evasion, resistance, escape (SERE), and airborne operations.

Physical Strengths

These human weapon systems have well-rounded whole-body strength and power. They can carry loads of over 100 lbs. (45 kg) over arduous terrain. The Arctic Infantryman can perform this type of physical work repeatedly with a little decrement in performance. However, the number of times an Arctic Infantryman can perform this type of work is highly individualistic, and the total length of time is still being determined.

Physical Limiters

These human weapon systems tend to have significant muscular and aerobic endurance limitations. This is not due to physical limitations but mainly to historical physical training. Sparta Science force plate data strongly suggests that the human weapon systems have an overdeveloped posterior chain compared to the anterior chain. Additionally, the data suggests that the human weapon systems have difficulty engaging core muscles through whole body compound movements. Finally, the human weapon systems have limited training to perform proper cold weather operations (i.e., cross-country skiing, snowmobile handling, etc.) and mountain warfare tactics.

Medical History

This is omitted due to Health Insurance Portability and Accountability Act of 1996 (HIPAA) regulations.

Conclusions

1. These human weapon systems are Arctic Infantryman with an average time in service 10.7 ± 6.2 years.
2. The human weapon systems have a strong strength and power background but have significant muscular and aerobic endurance limitations.

Key Questions

1. N/A.

Comparative Analysis

The comparative analysis allows the human weapon system or the HPO personnel to compare the information gleaned from the mission-oriented and human weapon system–oriented needs analyses. This comparison provides insights into how the human weapon system performs within the career field and mission set. Furthermore, this comparison can glean specific insights into the human weapon system's strengths and limiters. This information is then incorporated, which guides the development of the human weapon system's training plan. As a result, the training plan is tailored specifically to the human weapon system, reducing the chances for injury and increasing the chances for mission success.

Conclusions

The Arctic Infantryman does not have the training or experience for Arctic/cold weather operations. The priority of the physical training program needs to be on the development of the aerobic metabolic system and muscular endurance. Strength training needs to be focused on the muscle groups necessary for ruck marching in winter conditions and cross-country skiing. In addition, the Arctic Infantryman needs to be educated on proper endurance training/execution and cold weather nutrition and nutrition strategies.

Knowledge Gaps

1. How is the caloric deficit going to be fixed?
2. How will carbohydrate and protein limitations of the DLA MCW/LRP be addressed? Will off-the-shelf supplements be utilized?
3. How is hydration going to be addressed? Is there a best practice for keeping water from freezing? Will off-the-shelf supplements be utilized to address electrolyte and mineral requirements for the hydration strategy?
4. Can a glucose monitor measure blood glucose levels to eliminate glucose shortage as a source for decrements in thermoregulation?
5. What are the appropriate work-to-rest ratios in this environment to mitigate sleep deprivation issues on thermoregulation and mental acuity?

Appendix 1: Arctic-Specific Nutrition and Hydration Considerations and Solutions

Nutrition and Hydration Limitations

The unit exercise physiologist (EP) briefed the team on target nutrition goals before the training. Based on peer-reviewed literature and conversations with researchers from U.S. Army Research Institute of Environmental Medicine (USARIEM), the team needed to target 5400–6800 kcal/day based on the day's activities. The macronutrient content should consist of 55–60% carbohydrates, 20–25% protein, and 15–20% fats. The DLA agency provides 12 menus of "Meal, Cold Weather/Long Range Patrol (MCW/LRP)" meals for cold weather operations (www.dla.mil). The average MCW/LRP contains 1540 kcal (15% protein, 35% fat, and 50% carbohydrate). The DLA suggested three menu bags per person per day and provided for 4500 kcal necessary to replenish energy loss from exertion in extreme cold. This deficit was noted as being a minimum of 900–2300 kcal short of the recommended daily intake.

Beyond foodstuffs, the team's EP recommended that the team assume a sweat loss rate of 1.5 L/h during moderate to heavy exercise based on the planned loading of clothing, gear, and equipment. The suggested liquid intake composition consisted of water and support electrolyte packets from the MCW/LRPs.

During the course, the team was given the North Atlantic Treaty Organization (NATO) Arctic-specific Meal Ready to Eat (MREs). The MREs are titled "Ranson for arktiskt klimat". Each meal consisted of 4900–5100 kcal with an average macronutrient breakdown of 53% carbohydrates, 15% protein, and 30% fats. In addition, the meal contained dehydrated meals with vegetables, various packets of nuts, Nutella-like spread, electrolyte mixes, and hard candies (Figs. 10, 11, 12, 13, and 14). Even though these MREs were lower in the suggested protein content, they are superior in overall caloric content, MRE weight and quantity, and carbohydrate content. However, due to environmental conditions, the team was limited to food and water intake. Each team member ate about 60–80% of the NATO Arctic MRE daily. Based on subjective debriefs when the team returned, the average caloric intake was about 3200–3900 kcal/day. This reported caloric intake leaves a significant caloric deficit of 1500–2200 kcal, assuming the team targeted the minimum daily caloric intake goal of 5400 kcal. The team ate 2–3 organized meals during planned halts and relied on hard candies and other smaller, caloric-dense foods to snack in between stops. Furthermore, the dehydrated foods strain each member's water supply (Figs. 20, 21, 22, 23, and 24).

The team could not meet the liquid target intake of 1.5 L/h due to only carrying 2 L of water. Each team member had to boil snow to create their water supply at each planned stop (usually 2–3 per day with 4–6 h of movement between each stop).

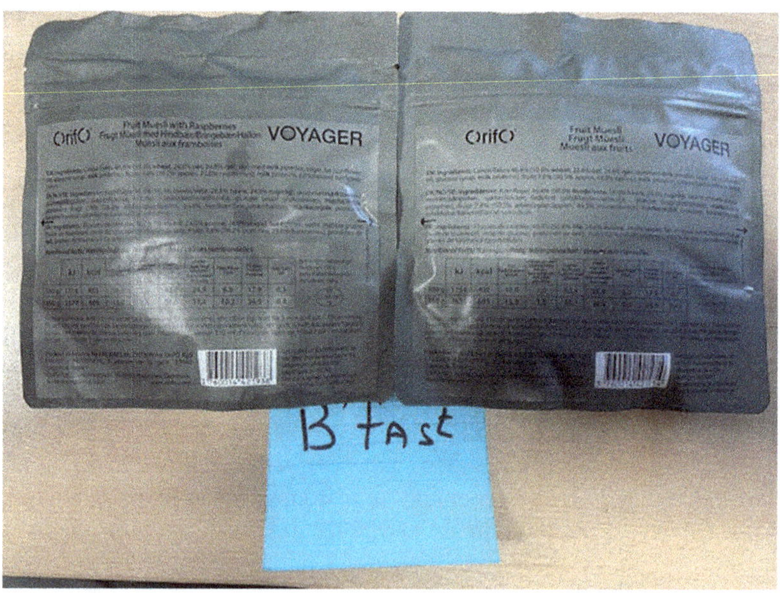

Fig. 20 This figure depicts an example of the breakfast options available in the "Ranson for artist klimat" meals. During the training, the team had only one breakfast option due to water limitations

Fig. 21 This figure depicts a lunch option with nutritional data in the Ranson for arktiskt klimat meal

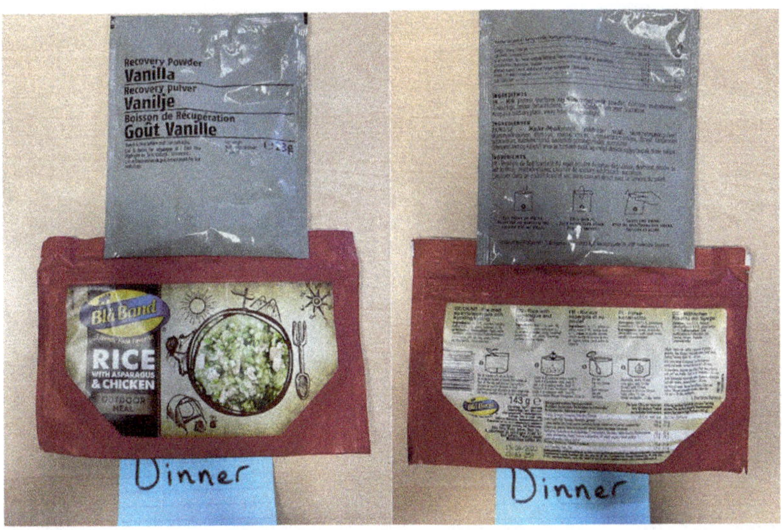

Fig. 22 This figure depicts a dinner option with nutritional data in the Ranson for arktiskt klimat meal

Fig. 23 This figure depicts the hydration options with nutritional data in the Ranson for arktiskt klimat meal

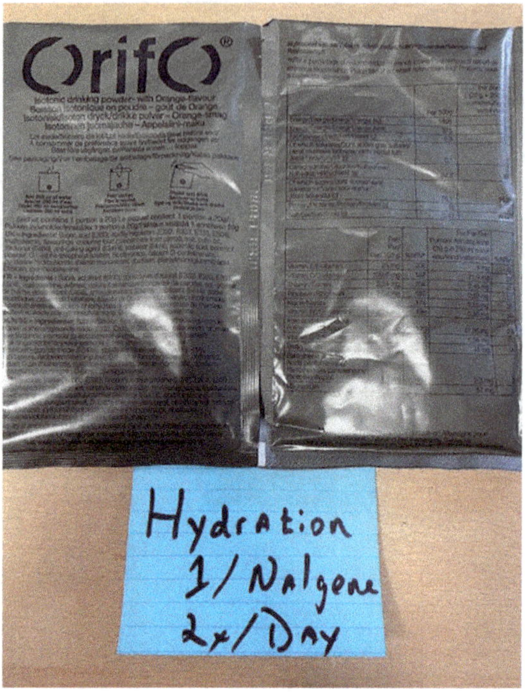

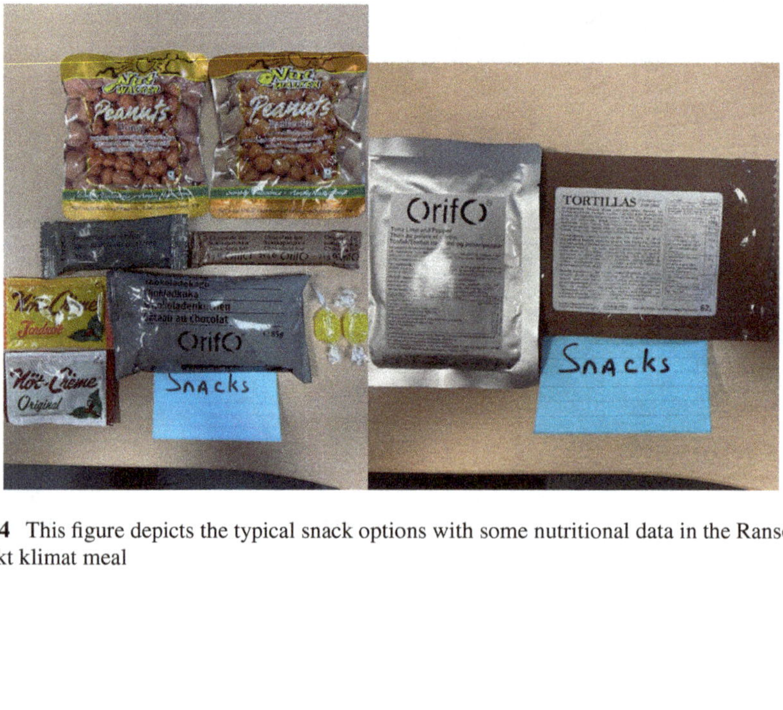

Fig. 24 This figure depicts the typical snack options with some nutritional data in the Ranson for arktiskt klimat meal

The dehydrated food could take up to 25% of each member's water supply (highly dependent on how the member prepared the food). The estimated water loss during each movement segment was about 6–9 L. With only 2 L of water, each member was short 4 L of water per segment. To compensate for this shortage, each member increased their night water intake as much as possible during the nightly stops.

The caloric and hydration shortages cause each member to lose weight. The average weight loss was 5–15 lbs. We could not measure other physiological metrics to see if these caloric and hydration deficits hindered performance. One aspect of the training was that each training segment ranged from 3 to 5 days with 1–3 days of recovery. These breaks in the school training calendar allowed the team members to recover. However, the given scenario would not be conducive to long-term operations for moderate to long durations (>5 days).

Conclusions

1. **Hydration.** Other hydration options need to be introduced into the Arctic Infantryman inventory. Other options allow for more water to be carried on the individual.
   (a) **Grayl 16.9 UltraPress purifier.** Several members of the permanently stationed units in Arctic area utilize this purifier during freezing weather operations (Fig. 25). These members have had significant success with this item, and its increased size allows more water to be carried in the individual's pack. The item costs $89.95 and can be found at https://grayl.com/collections/ultrapress-filter-purifier/products/16-9oz-ultrapress-filter-purifier-water-bottle-covert-edition?variant=39600436805711.

Fig. 25 This figure depicts the Grayl 16.9 oz. UltraPress purifier

Fig. 26 This figure depicts the 48 oz. Nalgene canteen

(b) **48 oz. Nalgene canteen.** This item was used by the team that participated in the Arctic Warfare School. This item survived the rigors of the course and did not leak (Fig. 26). The flexible canteen costs $19.32 and can be found at https://nalgene.com/product/48oz-nalgene-cantene/.

(c) **40F below-cold weather bottle**. This bottle is rated for -40F temperatures and can handle hot and cold liquids (Fig. 27). It comes in various sizes .5 L, 1 L, 1.5 L, and 2 L. The 1.5 L bottle costs $29.95 and can be found at https://40below.com/product/cold-weather-water-bottle-1-5-liter/.

2. **Nutrition**.

(a) **MREs.** The Ranson for arktiskt klimat is a superior product than the MCW/LRP provided by DLA. The Ranson for arktiskt klimat is superior because it provides more calories per package, a better percentage of macronutrients as identified in this needs analysis, and caloric-dense food that is easy to eat on the move (i.e., the candies). Future Arctic/extreme cold weather operations should consider utilizing these meals over the MCW/LRP meals in the DLA inventory.

(b) **Supplementation.** Due to environmental and movement limitations, not all food can be consumed. The caloric deficits seen during the Arctic Warfare training in January 2022 and in previous research are the primary cause of drops in body weight. Therefore, meal replacements and protein supplements in powder form should be considered in conjunction with the Ranson for artiskt klimat meals. These supplements can come in individual, single-serve sachets for each storage and use. These foodstuffs can provide a quick source of nutrition and increase protein intake to preserve muscle mass and body weight.

Fig. 27 This figure depicts the 40 F-be low water bottle

(i) The HPO nutritionist/dietitian can work with supplement manufacturers to create a blend that will work well with the target unit/population. These supplements can be purchased with Government Purchase Card (GPC) if used only for mission purposes.

(ii) The downside to these supplements is that they utilize water to be ingested. This could be an issue depending on how much water the individual may have available.

3. **Equipment**. One of the outputs of this working group was a suggested Unit Basic Load (UBL) for every Arctic Infantryman working in the Arctic/extreme cold weather environment. This equipment list with quantities and prices can be found in Attachment 1.

4. Figures 25, 26, and 27.

Appendix 2: Applicable Research Yet to Be Field Tested

Biomarkers for Safety and Performance

As stated earlier, the excessive cold puts a strain on the HWS. Cold stress can come from the environment or mission-related influences. These stressors impose substantial thermal strain that can cause decrements to work capacity and recovery [30]. The cold–stress response (i.e., vasoconstriction, shivering, etc.) is due to the

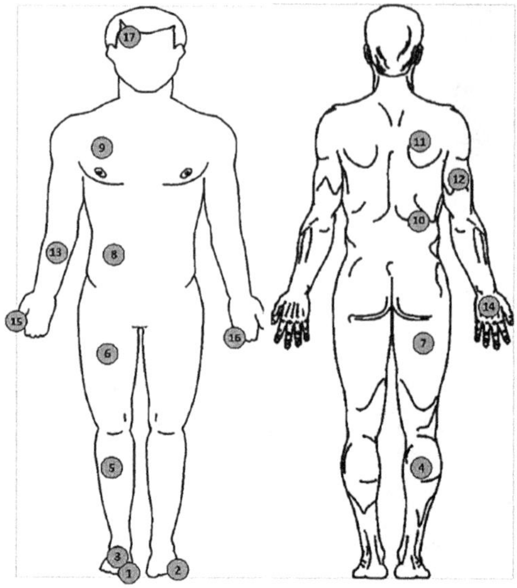

Thermistor placement

1. Big toe
2. Little toe
3. Foot
4. Calf
5. Shin
6. Thigh
7. Rear leg
8. Abdomen
9. Chest
10. Lower back
11. Upper back
12. Triceps
13. Forearm
14. Hand
15. Thumb
16. Little finger
17. Head

Fig. 28 Thermistor placement (T_{sk})

rate of heat loss. The way to monitor the severity of the cold–stress strain is via body temperature [30].

"Recognizing the stages and severity of cold strain can be achieved by assessing three main biomarkers: Skin temperature (T_{sk}) (peripheral and proximal), shivering thermogenesis, and core temperature (T_c)" [30]. The changes in the T_{sk} could be determined utilizing a calorimeter system and calculations (Fig. 28), as demonstrated by Campbell and Florian 2019 [30, 31].

Electromyography (EMG) and mechanomyography (MMG), which are considered the gold standard, should be used to measure shivering thermogenesis [30, 32]. The data from the EMGs and MMGs can calculate thermogenesis calculations as described in Arnold et al. 2020 [32]. Finally, T_c should be measured via a field expedient and reliable methodology in austere conditions. Based on literature and conversations with Air Force Research Lab (AFRL) and the Army Research Lab in Natick, MA, the best options for T_c in the field are the temperature pills or measuring the tympanic membrane.

Active Heating Methodology

During resting, cold-water immersions, arms, legs, and feet heating were the best region scheme to provide thermal protection and minimize battery/electrical power consumption [31]. However, the difference in the thermal conductivity of air and water may cause a significant difference in active heating methodologies. Additionally, movement versus rest may cause different regional heating necessities to obtain thermal balance. However, heating the arms, legs, and feet provides a good starting point since no published research could be found investigating optimal heating patterns while performing activities in extreme cold weather conditions.

References

1. Haff G, Triplett NT. Essentials of strength training and conditioning. 2014.
2. Myers C, Allen H. Triathlon Training with Power [Online]. 1st ed. Cognella Publishing. https://titles.cognella.com/triathlon-training-with-power-9781793532725 [11 Sep. 2021].
3. Tharion WJ, Lieberman HR, Montain SJ, Young AJ, Baker-Fulco CJ, Delany JP, Hoyt RW. Energy requirements of military personnel. Appetite. 2005;44:47–65. https://doi.org/10.1016/j.appet.2003.11.010.
4. Beck HE, Zimmermann NE, McVicar TR, Vergopolan N, Berg A, Wood EF. Present and future köppen-geiger climate classification maps at 1-km resolution. Sci Data. 2018;5 https://doi.org/10.1038/sdata.2018.214.
5. BM M, SJ C. Nutritional needs in cold and in high-altitude environments: applications for military personnel in field operations.
6. Marriott BM (Institute of M (US) C on MNR, Carlson SJI of M (US) C on MNR. Nutritional needs in cold and high-altitude environments. 1996.
7. Castellani JW, Spitz MG, Karis AJ, Martini S, Young AJ, Margolis LM, Phillip Karl J, Murphy NE, Xu X, Montain SJ, Bohn JA, Teien HK, Stenberg PH, Gundersen Y, Pasiakos SM. Cardiovascular and thermal strain during 3-4 days of a metabolically demanding cold-weather military operation. Extrem Physiol Med. 2017;6:1–14. https://doi.org/10.1186/s13728-017-0056-6.
8. Martineau L, Jacobs I. Muscle glycogen availability and temperature regulation in humans. J Appl Physiol. 1989;66:72–8. https://doi.org/10.1152/jappl.1989.66.1.72.
9. Jeukendrup AE. Periodized nutrition for athletes. Sports Med. 2017;47:51–63. https://doi.org/10.1007/s40279-017-0694-2.
10. Jones PJ, Jacobs I, Morris A, Ducharme MB. Adequacy of food rations in soldiers during an arctic exercise measured by doubly labeled water. J Appl Physiol (1985). 1993;75:1790–7. https://doi.org/10.1152/jappl.1993.75.4.1790.
11. Beals K, Perlsweig KA, Haubenstriker JE, Lovalekar M, Beck CP, Yount DL, Darnell ME, Allison K, Nindl BC. Energy deficiency during cold weather mountain training in NSW seal qualification students. Int J Sport Nutr Exerc Metab 29: 315–321, 2019. https://doi.org/10.1123/ijsnem.2018-0041.
12. Powers SK, Howley ET. Exercise physiology: theory and application to fitness and performance. 8th ed. New York: McGraw-Hill Humanities/Social Sciences/Languages; 2011.
13. Kenney WL, Wilmore JH, Costill DL. Physiology of sport and exercise. 7th ed. Human Kinetics; 2020.
14. Vallerand AL, Frim J, Kavanagh MF. Plasma glucose and insulin responses to oral and intravenous glucose in cold-exposed humans. J Appl Physiol. 1988;65:2395–9. https://doi.org/10.1152/jappl.1988.65.6.2395.

15. Gagnon DD, Perrier L, Dorman SC, Oddson B, Larivière C, Serresse O. Ambient temperature influences metabolic substrate oxidation curves during running and cycling in healthy men. Eur J Sport Sci. 2020;20:90–9. https://doi.org/10.1080/17461391.2019.1612949.
16. Jeukendrup AE, Gleeson M. Sport nutrition [Online]. https://www.amazon.de/Sport-Nutrition-Asker-Jeukendrup/dp/1492529036/ref=asc_df_1492529036/?tag=googshopde-21&linkCode=df0&hvadid=310619028235&hvpos=&hvnetw=g&hvrand=13307388785746463045&hvpone=&hvptwo=&hvqmt=&hvdev=c&hvdvcmdl=&hvlocint=&hvlocphy=9042720&hvtargid=pla-524500659240&psc=1&th=1&psc=1&tag=&ref=&adgrpid=65539873561&hvpone=&hvptwo=&hvadid=310619028235&hvpos=&hvnetw=g&hvrand=13307388785746463045&hvqmt=&hvdev=c&hvdvcmdl=&hvlocint=&hvlocphy=9042720&hvtargid=pla-524500659240 [12 Sept 2021].
17. Baker LB, Wolfe AS. Physiological mechanisms determining eccrine sweat composition. Springer Berlin Heidelberg.
18. Barnes KA, Anderson ML, Stofan JR, Dalrymple KJ, Reimel AJ, Roberts TJ, Randell RK, Ungaro CT, Baker LB. Normative data for sweating rate, sweat sodium concentration, and sweat sodium loss in athletes: An update and analysis by sport. J Sports Sci 37: 2356–2366, 2019. https://doi.org/10.1080/02640414.2019.1633159.
19. Haslam, D. R. (1980). The effects of sleep loss upon the motivation, morale, and mood of the soldier. In NATO symposium on Motivation and Morale in the NATO Forces, Brussels.
20. Shattuck NL, Matsangas P, Dahlman AS. Sleep and fatigue issues in military operations. In: Sleep and Combat-Related Post Traumatic Stress Disorder. 2017.
21. Aidman E, Jackson SA, Kleitman S. Effects of sleep deprivation on executive functioning, cognitive abilities, metacognitive confidence, and decision making. Appl Cogn Psychol 33, 2019. https://doi.org/10.1002/acp.3463.
22. Sun J, Wu J, Hua F, Chen Y, Zhan F, Xu G. Sleep Deprivation Induces Cognitive Impairment by Increasing Blood-Brain Barrier Permeability via CD44. Front Neurol 11, 2020. https://doi.org/10.3389/fneur.2020.563916.
23. Goel N, Rao H, Durmer JS, Dinges DF. Neurocognitive consequences of sleep deprivation. Semin Neurol 29: 2009.
24. Lamon S, Morabito A, Arentson-Lantz E, Knowles O, Vincent GE, Condo D, Alexander SE, Garnham A,Paddon-Jones D, Aisbett B. The effect of acute sleep deprivation on skeletal muscle protein synthesis and the hormonal environment. Physiol Rep 9, 2021. https://doi.org/10.14814/phy2.14660.
25. Goldberg AL, Tischler M, DeMartino G, Griffin G. Hormonal regulation of protein degradation and synthesis in skeletal muscle. Fed Proc 39, 1980.
26. Gopalakrishnan A, Li ;, Ji L, Chiara Cirelli ; Sleep Deprivation and Cellular Responses to Oxidative Stress [Online]. https://academic.oup.com/sleep/article/27/1/27/2707929.
27. Dáttilo M, Antunes HKM, Galbes NMN, Mônico-Neto M, De Sá Souza H, Dos Santos Quaresma MVL, Lee KS, Ugrinowitsch C, Tufik S, De Mello MT. Effects of Sleep Deprivation on Acute Skeletal Muscle Recovery after Exercise. Med Sci Sports Exerc 52, 2020. https://doi.org/10.1249/MSS.0000000000002137.
28. Li JB, Goldberg AL. Effects of food deprivation on protein synthesis and degradation in rat skeletal muscles. American Journal of Physiology 231, 1976. https://doi.org/10.1152/ajplegacy.1976.231.2.441.
29. Kolka MA, Martin BJ, Elizondo RS. Exercise in a cold environment after sleep deprivation. Eur J Appl Physiol Occup Physiol. 1984;53:282–5. https://doi.org/10.1007/BF00776603.
30. Lee JKW, Tan B, Kingma BRM, Haman F, Epstein Y. Biomarkers for warfighter safety and performance in hot and cold environments. J Sci Med Sport. 2022; https://doi.org/10.1016/j.jsams.2022.12.006.
31. Campbell JE. Measurement of regional heat exchange using direct regional calorimetry in resting subjects immersed in cold water. 2019.
32. Arnold JT, Hemsley Z, Hodder SG, Havenith G, Lloyd AB. Reliability and validity of methods in the assessment of cold-induced shivering thermogenesis. Eur J Appl Physiol. 2020;120 https://doi.org/10.1007/s00421-019-04288-2.
33. Rice SM, Purcell R, De Silva S, Mawren D, McGorry PD, Parker AG. The Mental Health of Elite Athletes: A Narrative Systematic Review. Sports Medicine 46: 2016.

Index

C. Myers, *The Human Weapon System*,
https://doi.org/10.1007/978-3-031-45061-7

www.ingramcontent.com/pod-product-compliance
Lightning Source LLC
Chambersburg PA
CBHW061639030826
49219CB00003B/36
9783031450600